WALLS OF GLASS ON A MOUNTAINTOP

My Six Years as a Fire Lookout.

By Alice A. Johnson

Alice A. Johnson

Jelm Mountain Publications
Laramie, Wyoming

WALLS OF GLASS ON A MOUNTAINTOP

by Alice A. Johnson

ISBN 0-936204-68-0

Library of Congress Cataloging-in-Publication Data

Johnson, Alice A., 1903-

Walls of Glass on a Mountaintop: My six years as a fire lookout. / by Alice A. Johnson.

p. cm.

ISBN 0-936204-68-0

1. Johnson, Alice A., 1903- . 2. Fire lookouts — United States — Biography. I. Title

SD421.25.J64A3 1989

634.9'618'0924—dc 19 89-1652

CIP

Photo illustrations by the author.

Front cover – Pelican Butte Lookout Tower covered with rime (a white icy coating from atmospheric moisture).

Design by James Beasley.

Published by Jelm Moutain Press,

Laramie, Wy. 82070.

Special Thanks

Dedicated
to my family and friends

* * * * * *

Special thanks to Nikki Elkind of Ft. Collins, Colorado for her criticism and suggestions, and for many hours spent in typing drafts.

I cherish the pleasant days we spent in hiking the beautiful, and often times strenuous, mountain trails of Colorado and Wyoming. Her encouragement and quiet sense of humor kept me moving on to reach my goal, be it the completion of this manuscript of memories, or the completion of an ardous climb to reach an alpine lake.

Table of Contents

FOREWORD

The lookout system to aid in fire detection was set up by the Forest Service early in the twentieth century. Shelters were constructed on high points where the firewatcher would have a panoramic view of the surrounding terrain.

Many times while spinning along a mountain road I would look up with a sense of intrigue and wonder at these tiny specks that dotted the skyline here and there. I secretly envied the occupants of those lofty cubicles, but other responsibilities required that I walk on lower ground. It was not until Old Father Time had my name on his list that the door to this venture was open to me. I made my escape from the stream of life that would have taken me over retirement hill, and for six years I was out of touch with Father Time.

Although all towers have similarities and may look alike from a distance, there is great variation in working conditions for the occupant, depending much upon the geographical location, height of the mountain, and height of the tower, as well as on the personalities and working methods of the supervisory personnel.

Every locale has its own operating procedures according to the needs of that area. To be a lookout, one must have the ability to cope with extremes, whether borne on the winds of a mountain or on the whims of his fellow men.

Old-time lookouts of Pelican Butte would have thought my way of living pretty easy with a butane range and heater, but I was content to enjoy the warmth of butane heat when high winds blasted the ice-crusted tower. I was also content that I did not have to heat my cook stove with wood, as old-time lookouts did, when temperatures hovered around one hundred degrees and the mercury hit the top of the thermometer inside the sun-blazed walls of glass.

This story relates only to my own experiences and observations and to my personal reactions to the life and events at these particular lookout stations during the seasons of 1965 through 1970, and may have little similarity to the life and experiences of other lookout personnel.

To make my story more readily understood, I have identified by a given name rather than the code that is always used in radio communication with the lookouts, ranger stations, and guard stations, as well as all cars and equipment used by the Forest Service. All stations and equipment have code names and numbers. Anyone occupying a station or operating equipment carries the code assigned to that station or equipment.

LOOKOUT — the word has a double meaning. It refers to the tower or station and also to the person manning the station.

Walls of Glass
on a
Mountaintop

Chapter 1

Price's Peak, 1965–66

I was half past sixty-one, near that time in life when one steps over the hill into the golden age of retirement — sixty-one, sixty-two, Social Security, here I come. I would fold away my unfulfilled dreams and see what retirement held for me. Then fate stepped in and gave my life a turnabout. A dream I had held for many years was now to be fulfilled — my dream of work as a forest lookout. I would live alone on a high mountain tower. As a lover of nature I would hope now to have a part in its protection.

It was quite by chance that I met Mr. and Mrs. Wain who had recently retired after eight summers as forest lookouts with the Department of Natural Resources in Washington state. Through Mr. Wain's recommendation I was chosen for his replacement. It seemed too good to be true. However, I was to learn over the next six years that reality does not necessarily follow the pattern of a dream. I would experience many highs and lows. There were drab days that seemed never-ending, sensational storms and tense situations in line with my work.

Mr. Wain was helpful in giving me information regarding the work, as well as timely tips on the quirks of nature, including human nature, that I would be likely to come in contact with.

Price's Peak in the Coast Range, thirty miles from the Pacific Ocean was to be my summer home for the next two fire seasons. I felt somewhat deflated when informed that my mountain was only 730 feet above the level of the sea. Not very impressive, I thought. It sounded much less impressive when a relative from Cheyenne, Wyoming said, "Do you know, if you were at that elevation here in Cheyenne you would be in a hole in the ground well over a mile deep?" Well, never mind, Price's Peak is higher than a hole in the ground, and it is higher than the country surrounding it. In spite of Cheyenne's lofty height of over 6000 feet that fair city is a hole in the ground compared to some of the country surrounding it. High or low is relative, or so I thought at the time.

I did not know what the season would bring, so preparedness was the word. I left Salem, Oregon, starting out on my new venture with a good supply of groceries, home-canned meat, fruit, vegetables and supplies for making bread. I brought a minimum of cooking utensils and dishes. I brought clothing and bedding suitable for any weather, books, correspondence supplies, my camera and a flashlight. I had tools that a woman could fix or repair anything with: hammer, saw, pliers, screwdriver, shovel and an axe. Besides these basic essentials of life, I had a few things that "might come in handy," such as rope, string, tacks, nails, Scotch tape and old rags, to name a few. With these supplies one could improvise anything.* I was ready now to step off into the wide blue sky.

I arrived at the headquarters in Montesano on June first, 1965, where I met my supervisors, the office crew, and Jack, the district warden. I followed Jack up to the lookout in his trail of dust, through several big mudholes and numerous locked gates. Jack unlocked and opened the gates. After driving through, not wanting to appear helpless, I jumped out of my car, then closed and locked the gates behind us.

How excited I felt as we rounded a switchback and the tower came into view, its massive frame silhouetted against the sky. I had arrived! Then the work began — all these supplies must be pulled to the top of this forty-foot structure. Jack hoisted my supplies, along with the firefinder, radio, and other working supplies, using a rope and pulley that hung from an arm on the upper part of the tower. I steered the

* All articles mentioned were essential to my work and way of life during my stint on the lookouts.

pulley arm over the rail and lowered the supplies to the catwalk surrounding the tower cabin.

Jack was not a man to waste time talking so, after a quick rundown on the use of the firefinder and radio, he took off and I was on my own to settle in and get acquainted with my new line of work.

My first job was sweeping up glass. Although the Wains had left the place neat and clean, vandals had shot out several windows during the winter vacancy. The panes were replaced before I arrived and some of the glass had been picked up, but glass still covered the mattress, the stove, and the floor. What a mess to clean up! Deer that had been tamed by Mrs. Wain were now dead, their legs laid out in a row beneath the tower. It was a sad lesson, but one I had to learn — don't make pets of the wild animals or some dastardly human might come up and kill the trusting creatures.

A district administrator arrived the following day to check my knowledge of topography maps and to make sure I could read and understand legal land descriptions. He explained the drainage systems of the area and said I would have trouble at first in judging distance from this station as the hills were heavily forested with no distinguishing landmarks. The administrator emphasized that radio communication was of equal value to the Forest Service as the work of spotting fires. The messages were coded as far as was practical. One radio served two channels.

Jack, the warden kept me supplied with water during the early part of the season. He came up once a week with twenty gallons and, in addition, my ever-welcome mail. With the twenty gallons I did my laundry, took daily baths, mopped the floor every day, kept the glass walls of the cabin clean inside and out, and had plenty left over for other household needs, although some water was used a second time. Water once used for rinsing dishes could be used again to mop the floor, and water in which I rinsed my clothes could be saved for the next wash or used for bathing.

I later learned of a good spring that flowed from the hill a mile down the road. I then offered to haul my own water as I needed the exercise and was glad to have that free time off the tower. By using a hose and funnel, I was able to fill the ten-gallon cans without removing them from the car, thus sparing myself the work of reloading the cans which weighed more than 100 pounds when full. By using the rope and pulley, I would pull the cans up onto the catwalk with the car. It took a lot of maneuvering and quick handling of the car to get the cans over the rail and down to the walk on the other side while the cans were twirling around on the end of a rope forty feet up in the air, with no one to swing the arm of the pulley.

From my Journal: The tower sits on a grassy knoll edged with foxglove in many shades of red to pink, pure white and lavender. They sometimes reach above my

Ten gallon water cans were hoisted to the cabin by car, using a rope and pulley. It took skill and quick action of the car to get the cans over the rail and down to the catwalk, while the cans were twirling and swinging on the end of the rope forty feet up in the air.

head to heights of six or eight feet. The surrounding hills are a rough sea of young hemlock and other evergreen and deciduous trees. The cabin is fourteen by fourteen feet square, a live-in tower with a catwalk two feet wide around the cabin. Inside the cabin are built-ins two feet high along one side. The firefinder stand is centered in the middle of the room. It also serves as a stand for the communication radios. Oh yes, I must not forget the little stool with insulated legs to scare off the lightning.

. . . The birds were the first to greet me. They perch on the arm of the pulley, looking in at me as they chatter away to each other, knowing they have nothing to fear for they have me in a cage. I envy them their freedom. They can fly the skies all summer while I am confined to the cage.

. . . The first few nights on the hill seemed spooky, everything so quiet except for the mournful sound of the wind around the tower. I often hear coyotes in the evening and morning hours. They sound very close; their screaming howl, distressed and mournful, rings sharply through the forest. A buzzard sails around the tower periodically. Is he checking me for a meal? I have heard that a lookout died at his post on one of the hills around here. He was overcome by gas from one of the appliances. I was told I should not use the floor furnace as it was not working properly and might not be safe. "You can turn the oven on for heat." I tried that, but it turned out to be unsatisfactory as fumes from improper gas combustion burned my eyes and throat, so I took a screwdriver and, applying my Irish ingenuity, I fixed the furnace. It has given no more trouble.

. . . I love to watch the storms coming in from the coast, sometimes striking the tower with force, then racing across the hills leaving a trail of fog devils streaming up from the ravines.

. . . Olga, an ex-lookout, publishes a monthly bulletin for and about the forest and the personnel. It is her contribution to keeping all "her people" in touch with each other. Her writing is spiced with humor, true tales, and sometimes a note of sadness. It is a way of getting acquainted with the voices we hear on the air, from co-workers we may never meet. I watch the mails for every issue. "Ten-eight" seems an appropriate title, as this is the code for signing on the air. The lookouts are asked to write in, telling something about themselves and any interesting experiences they may have had as lookouts. I sent in the following soon after the start of the first season:

Hello from Price's Peak: You asked for lookout experiences. Well, there has not been much excitement on Price's Peak, but I almost created some. I awoke about midnight. Fog hanging low in the valley reflecting light from a bright moon gave a weird, silvery glow to the night. As the moon lowered into the fog, it became magnified and turned an orange-red, bearing no resemblance to the moon. The spectacle would have been frightening had I not seen it in the "moon stage." As it settled deeper into the fog, it looked for all the world like a fire. If I had not seen it before this stage, I might have turned it in for a fire. Wouldn't I be popular with the wardens if I had gotten them out of bed at midnight to chase the moon?

Olga's reply: "Chasing the moon, I do believe would be something new. But we have had our long-suffering wardens running after trains — searching for ships — and hunting sunlit sprinklers at — all — hours."

. . . I had a caller today, a state-employed bear hunter. He had found a bear and killed it, but lost his dogs in the process. A few bears are turned loose wearing radio transmitters on their collars; these emit signals so the bears' movements can be traced and their habits studied. Bears are smarter than people; they have no radios on us, but believe me, our movements are traced and they know our habits. It took two men and four dogs to catch and kill this bear. The hunters are hired by the state as it is thought by some that bears are killing the trees; some feel otherwise. There are two factions, the "Save our trees" (S.O.T.'s) versus the "Save our bears" (S.O.B.'s). A bear festival is held once a year in the logging town of McCleary. Here the S.O.T.'s gather to eat bear stew and have a high old time. A few of the lookouts are granted time to come down off their mountains and gorge on bear stew, while the rest of us sit on our towers and stew, because we are the S.O.B.'s.

. . . My work here is interesting. I may not know what is going on in the world and outer space, but I know what is going on in the woods of western Washington and that is interesting enough. It stays light here until ten o'clock; although not on duty officially at that hour, I am on alert from daylight till dark and may not leave the station without permission, either day or night. I awaken every morning in a gray fog (literally, that is). Some mornings I cannot see a thing beyond the fog-bound tower. It is a strange feeling to be wrapped in a cloud. As morning wears on, the trees become visible and by noon it has usually cleared away.

On my return trip from Montesano last night, I came upon a doe in the middle of the road. At her feet lay a tiny fawn, hugging the ground as fawns will in the first few days of life. Nothing will rouse a fawn when in this flattened position; they seem glued to the ground. This is nature's instinct for protection. However, the middle of a forest road seemed an unwise place to hide. The mother made no move and, not having room to pass, I got out of the car and drove her away. I picked up the fawn and carried it off to one side. It was probably no more than a few hours old, as it had no instinct to follow its mother. The little thing followed me back to the car after my every attempt to remove it. I had to carry it up the roadside bank where its mother, in wide-eyed terror, was watching her baby being held in the arms of a monster.

. . . Last night I had a scare. This is not a bear story. Well, not the conventional kind, anyway. I was the "bare." I was ready to take my bath, in the old-fashioned way, of course. I pulled my "drip-and-shiver" bathmat out from the head of my bed where I keep it out of sight when not in use. Sprawled on that mat was the most hideous thing I ever did see. It looked like a cross between a crab and a spider. It was about two inches long, I do believe, and keeps getting longer as I think of it crawling only a few inches from my head while I slept, perhaps ever since I arrived. It nearly threw me into a frenzy. I now had to face the problem of getting rid of it. Well, believe me, the tower was in a tizzy until I got that crab-spider licked. Why did I so dislike that creature when I welcome the deer? The spider was much more colorful.

I do see a few things up here besides bugs, bear hunters and deer. I see the snow-capped Olympics and Mt. Rainier, which is too far away to be very spectacular. The hemlock trees are beautiful at this time of year, a sea of green. Sometimes when I am out of sorts they seem like a big bowl of spinach garnished on top with a hard-boiled egg. That's me!

Only once have I seen the sun rise, as it has been so foggy in the mornings. This fog is the wettest stuff I have seen except for the ocean. Envelopes seal themselves, and my clothes feel damp when I put them on.

I have not used my little transistor radio much. I would rather listen to the "whoo-whoo" of the owls and musical notes of the songbirds. I sat on the catwalk for over an hour last night, listening to a chorus of coyote pups. I do not know what their trouble was, but they kept up their little yippy-yaps longer than I have ever heard them before.

If I do not get my letters written, I cannot give the usual excuse, "I was just too busy," or "time goes so fast." Last week I had very little to do, and nothing goes slower than time around here when I am caged alone on the lookout tower. I hope to have two days off soon. I am on duty fourteen days, then have two days off, unless fire danger dictates otherwise. I had my first fire last night, just a small brush fire.

One thing I have learned — never take a bath unless you have a pencil in your teeth, a scratch pad under your arm, and a towel around your neck. I was called while taking my bath the other night, and tangled the code in answering the message. You might say the message was all wet.

. . . A few days ago I heard the sound of a car that must have been stuck in that mudhole down there. The mudhole is half a mile down, on a road that branches from the lookout road. I heard the racing of a motor and a call for help. Calling to headquarters, I asked for permission to leave the tower long enough to find out what was going on down there. Someone surely must be hurt. I was told I should not go down. "If someone is stuck you will only get stuck, too," came the reply. I knew the mudhole was not big enough to hold both of us, but could not explain this to the dispatcher, and I could not leave the station without permission, much as I wanted to. The roar of a motor and calls for help continued periodically. I could stand it no longer, so I called down to the office again. Still they did not think it safe for me to go down. This had been going on for five hours when I heard the vehicle pull up over a rise. Driving slowly and headed for town, he passed under the brow of the peak. Due to the forest and a switch in the road, he was never in sight from the tower.

The next morning I took my usual walk and found many cougar, bobcat, and bear tracks in the mud. The only man tracks I saw told me the driver had ventured out of his car once, placed a small branch under one rear wheel, then got back in the car again. The animal tracks were all around where the car had been stuck. I cannot imagine what man would sit in a mudhole all afternoon, out in the middle of nowhere, spinning his wheels and yelling for help without making more effort than that to get out.

I often go down in the morning to eat a few salal berries and to see what has been going on in the animal world. I always see tracks of some kind, bear, cougar, bobcat, deer, and coyote, besides the smaller animals. I have nothing to fear, as we have our times allotted and seldom meet. The early morning hours are mine; they have the night. However, I do carry a small axe just to make myself feel braver.

The head lookout called this evening to say she was sure the cries I heard were those of a cougar. I know there is a cougar in the area, and they do have the cry of a terrified human which makes your blood run cold. I have also heard their sound like that of a baby crying. But I am sure this was a man, as I so clearly heard the words, "Help! Help!" and I do know a car was stuck down there from noon until five in the afternoon. He may have seen a cougar or perhaps the tracks of animals and was afraid to get out of the car. He was evidently someone unused to the forest roads as his tires were bald. I could see the tracks where his car had been slipping even before he had reached the mudhole. The moisture was from fog as we have not had rain.

. . . I had a round with an army of hornets that built a huge paper nest beneath the lower beam of the tower. The nest was directly under the steps. The hornets chased me up the tower, buzzing angrily about my head, making me very nervous. I had to do something about that! One cold, windy night after the hornets and I had both signed off for the night, I covered the nest with a large plastic sack and cut a hole just large enough to admit the tip of an "insect killer" spray can. The roar of wings inside that nest was horrible. I dashed up the steps, feeling like an escaped murderer, but I could not escape that roar of death. I was not tormented with hornets again.

. . . I had a little excitement the other night when I heard a cry that had the sound of a terrified deer. I went down the next morning to read the animal tracks. I was able to reconstruct the scene where a cougar had jumped from a stump and seized a fawn. The mother deer had evidently jumped down from the bank above. Their heavy tracks in the moist soil, along with the tiny tracks of the fawn, told the story of the cry in the night.

. . . What a beautiful morning. Beneath me lies a soft, white mist, tinged with an aura of blue light. The cool, bright air is silent but for the happy spring song of many birds that flit past the tower, diving in and out of that shimmering sea of sunlit fog. The only human voice on the air was the early morning check call of the lookouts. I turned on the transistor radio to a church service in

Aberdeen, listened to gospel singing and heard the minister speak of living above the cares of this life in the light of God's love. The beauty of the sermon seemed very real in the light of this atmosphere, literally separated from earth by the silent clouds, surrounded by sky to blue infinity.

. . . I awoke to the usual chorus of bird song, but was surprised when I looked out and saw no earth below me. Had the tower come loose from the earth? Was I sailing through the clouds No, just a little more fog than usual, and the wind, sweeping fog past the glass walls of the tower cabin, gave a sensation of sailing.

. . . Last night I went down to the spring for drinking water, but found the beavers had a building project going on, making the water unfit to drink. I found a smaller spring farther up the ravine. After digging it out, I filled the cans with a bucket.

. . . I stepped out on the catwalk in my night attire to snap a picture of an unusual sunset, dark gray cumulus above the horizon against a vivid orange sky. I heard a plane, or was it a tractor, working down below? The roar was getting closer and louder as a helicopter rose up from the gulch and headed straight for the tower. In disgust I thought, "If the windows were open he would have flown right in. I'd like to swat him with my flyswatter." He appeared again a few days later, touching down between the tower and the garbage pit. I thought he would surely clip the roof of the tower with his whirling blades, but he landed as neat as a bird on a twig. He politely called for permission to come up the tower and we had a pleasant visit. Visitors are always welcome up here; well, almost always.

. . . Two very tough-appearing young men on red motorcycles rode up to the peak today. I did not go out to greet them, but watched through a hole in the floor until they started up the tower steps. Not knowing what to expect, I stepped out onto the catwalk to see the trapdoor swing up and a face with mangy black hair pop up through the opening.

I said, "Maybe you boys had better run on down the stairs."

The mangy black-haired face just glared at me, making no move.

The second boy asked, "Can't we come up?"

I replied more pleasantly than I felt, "I'd rather you didn't."

The first boy, still with that horrible stare, then said in a sarcastic manner, "OK, then we'll go down."

I was grateful that, just as they were coming up, one of the lookouts, a man with a deep voice, had been on the air. Perhaps they had heard him and thought he was with me. Really, all the protection I had was a big hammer and a downhill swing. I did not take my walk tonight; I felt too nervous. I cannot get those mean-looking faces out of my mind. I went down from the tower long enough to saw off a piece from a two-by-four that was lying beneath the tower, then fitted it between the trapdoor and the roof overhang. I will feel much safer now. If anyone should try to climb over the rail of the catwalk, which I have heard of them doing, I will hit their hands with my hammer and leave the rest to gravity.

. . . Two fires today, neither of them on my area. Five acres burned on the Bear Gulch fire.

. . . Very foggy and windy this morning. It looks like a blizzard, almost, as the fog sails over the treetops. The fire crew is still on the Bear Gulch fire. With everything dripping with fog up here, it is hard to visualize the fire hazard down below.

. . . Windy and cloudy again. Today I repaired the bed, using my hydraulic jack to unbuckle the warped sheet of plywood beneath the sagging mattress. I then cut a few sticks from a cascara tree to prop up the plywood. I can now open the drawers beneath, where I may store my clothing.

. . . I was awakened in the night by an intense glow from a full moon reflecting its light on a layer of fog drifting among the treetops. I slipped onto the catwalk. The night was steeped in an eerie fusion of darkness and light. Creeping webs of moonlit fog intertwined the tips of the hemlock trees. The tower shadow, like a finger of darkness reaching across the veil of fog, was encircled at the very tip by a halo of pure, white light, adding a spectral dimension to the night. I heard a soft tapping down in the trees, like raindrops falling on the broad-leafed maples. Could it be raining beneath those clouds? The night was too impressive to leave, but I must have sleep to be alert for work in the morning. Back in the cabin, I thumbtacked newspapers around my bed to shut out the awesome light of the night.

. . . This morning the sun rose brightly over a sea of clouds. Its rays reaching across the sky cast a rosy glow to the mist rising from acres of clouds that surrounded the peak.

. . . The foxglove has been beautiful all summer. The blossoms open a little further up the stalk each day, marking time on the passing season.

. . . It is foggy and dark this evening. Down in the timber the coyotes are howling; their plaintive wails seem to split the air. Distress? Despair? Whatever, it's a forlorn sound within this dark and dismal atmosphere.

. . . Although it was cool this morning, the weather report was "hot and dry." By afternoon the thermometer read 106 degrees. The sun reflecting through the walls of glass made the heat seem unbearable. I pried open a window that hung on rusted hinges; then with rope and eye bolt I tied it up to the roof overhang, but no breath of air was stirring.

. . . It is 109 today. Even the wooden bench burns my skin, and I do mean skin, for I cannot bear the heat of clothing against my body. I'm wearing small towels pinned over my hips, hoping for a quiet day with no company. And quiet it is. The air is dead, all life crushed in its oppression. No voice on the air but the sign-on and sign-off of the lookouts and wardens. No bird song or cry of the coyotes, and, thank goodness, no fires. I put the butter and cheese in a gallon glass jar and buried it under the trees, but still it melted and ran together. Heat waves wrinkle the air. There is no brilliance to earth or sky. Have they, too, melted and run together?

. . . The tall foxgloves are now bending with age. The blossoms, having reached the very tips, cling to the aging stalks; reluctant they seem to release their hold on life. I am now counting the days until I go down for the season.

. . . "Price's Peak — 2-18": The coded call saying, "Have the coffeepot on; I'll be there in fifteen minutes, more or less." The call came from the man who grades the forest roads. He had been working the road to Price's Peak all day. I had not met this man, but had heard it said, "He is always getting his grader stuck and calls on the radio for help in a voice so loud it fairly shakes the timber. He is the biggest and roughest-looking man you ever saw." Minutes after the call he pulled his roaring machine to a stop at the base of the tower. What a monstrous piece of equipment it was, a relic of better days. Like a dinosaur, its long iron neck stretched over the peak, the bladed head hanging down off the faraway end. What a genius, I thought, to maneuver this thing on these steep, single-track forest roads! A few cups of coffee and sandwiches later, I had formed my own opinion of this big man. He was looking forward to retirement when he and his wife could spend their time rock-hunting, cutting and polishing gems and making jewelry. He had a few pieces with him, his own make and design. I marvelled at the delicate beauty fashioned by those huge hands. I admired the gentle nature of this rough and dusty giant.

. . . It was mid-September when the heavy rains came. The air was filled with water and the roads were streams of mud. The lookouts received their calls to come down for the season. I worked frantically all morning, lowering my belongings to the ground on the end of a rope. I then carried down what I could on my back, picked up what I had lowered on the rope and loaded it all into the car, repeating this many times until everything was down and ready to go. One suitcase broke loose from the rope when halfway down and plopped into the mud twenty feet below. Fortunately, it did not break open. In the middle of all this, a call came from the grader man: "The grader has a flat." I relayed the message to headquarters. The only reply was, "10-4," a mere acknowledgement of the call. I received several calls regarding the flat; that man on the grader was getting impatient. All I could do was relay his messages on.

I was soaking wet when I finished loading the car. Since all my clothes were packed, I stripped off my outer garments, strung a rope from the firefinder stand to a leg of a table and hung them over the furnace to dry. When they were barely half dry, a call came from headquarters. The assistant administrator had left there some time ago.

"Have you heard anything from him?"

"Negative," I replied.

Thank goodness, I thought, as I jumped back into my wet clothes and tore the line off the stand, as it is strictly against the law around here to attach anything to or lay anything on the firefinder stand. Then I waited and waited for that truck. Finally it came, grinding up the hill in lowest gear. The warden, Jack, was with the assistant administrator. Having been stuck several times, they had pulled themselves out with the winch. I called down to the assistant as soon as the truck pulled up by the tower and told him about the trouble the grader man was having.

I expected, of course, that he would rush right off to his aid. Instead, he came up on the tower and sat for an hour and a half, eating lunch and telling stories of the movies that had been filmed in this area. The stories were interesting enough, really, but I kept thinking of the grader man sitting out in the cold rain all those hours. It was popular and expedient, perhaps, to like this assistant administrator. He was a nice young man. Like fog in the valley on a bright spring morning, he was on the rise. One could hope to get caught in the updraft.

I wanted to take my car down, simply to let the men know I could, but Jack said, in his quiet way, that he did not want me to drive down those roads by myself. He drove my car over the worst of it, then caught a ride with the assistant administrator who was taking a branch road down to the grader. As Jack said good-bye he shook my hand and said "I'll see you next year." However, I was never to see this kindly man again. A week later he was found dead, out in the timber where he had been working. It was a shock to all who knew him. I could only wonder, had I still been on duty, might he have called, might I have heard him and perhaps saved his life? It was not to be. Mr. Wain had said, "Jack is a good man; he will help you all he can." I found this to be true.

* * *

I was on Price's Peak for two fire seasons. Here I lived with nature, yet had contact with the human race. The work of a firewatch is interesting, but demanding, and calls for deep dedication, for the hours are long and the pay is short.

When the station was closed at the end of my second season, the eyes of Price's Peak were closed forever. Observation planes became the fire spotters, replacing many devoted servants of the Forest Service. A few stations were left open, largely for the purpose of communication.

Chapter 2

Lenhart Butte & Lookout Mountain – 1967

Lenhart Butte Tower. Lower part of tower is not shown here due to the height of the trees.

I started my third year as forest lookout with a state-funded fire protection association on Lenhart Butte with headquarters at Molalla, Oregon.

This 2,000-foot, forested knoll in the Cascade Mountain foothills had easy access to civilization. Here I learned that lookout towers, like forest lookouts, may be any size or age. This 110-foot tower showed the ravages of time and unfriendly elements. Carpenter ants were busy in the old wooden legs of the tower and specks of wood dust piled up on the supporting brace below. When the pile grew big enough to be noticed, a crew of ants was dispatched to kick the wood dust into the tall grass below, out of sight of nosy lookouts. One leg of the tower was honeycombed to the consistency of Swiss cheese.

The cabin on top of the tower was less than six feet square with a foot-wide walkway around the outside to stand on when washing windows. It had a trap-door over a crawl-through space in the floor, and a narrow shelf for work papers and maps. The firefinder stand was evidently built for a tall man and could not be lowered. In order to read the map on the firefinder, I had to stand on the lightning stool, which had legs centered in the middle. With any movement of my body it would tip over — very frustrating when trying to get a legal description on a smoke. It must have been built for a milking stool. I asked for topographic maps and was told the lookout usually furnished her own. However, I had no chance to get down to buy them.

A former lookout on this ancient tower had painted these words on a roof support: "Oh Lord, How Great Is Thy Universe," evidently a thought expressed by someone able to look above the crumbling, man-made things of earth.

From letters to friends: . . . My living place here is a musty old cabin called a ground house, so-called because it rests on the ground. It would be cozy enough had it not been in use as a winter refuge for a motley horde of bugs and insects, the dying, the dead, and those many reincarnated. How do you tell a million bugs that spring is in the air, it's time to go out and play?

An active army of hornets has its barracks in the rear wall, with an outside entrance to the back. Their hours of activity coincide with mine, but due to our separate entrances, we have little conflict.

An old range serves for heating and cooking. Wood is supplied in stove lengths, but needs to be split.

The work as lookout and weather observer is of first priority, but sandwiched in somewhere, between daylight and dark, I have managed to get most of the bugs and dust swept off the bed and a shelf or two for my supplies. The range was filled from ashpan to lids with ashes. The shovel I carry for emergency was too large for this job, so I used my hands for an ash shovel until the warden came up and finished the job for me.

Dark, dank and dismal as the cabin is inside, one has only to step beyond that grimy door to enjoy the radiant light of spring. The sky is wide, and alive with chittering notes and flitting wings of myriad birds. Wild flowers flash their dots of color amid the tall green spears of grass. A shy little cottontail skitters about near the brush at the edge of the timber.

My work as weather observer, as well as other ground work and meals, requires many trips up and down that tall tower each day. I rush through my work on the ground for, though surrounded by natural beauty, there is no distant view, which is very frustrating to a lookout whose duty is watching for fires.

* * *

I was at this station long enough to enjoy a visit from a few relatives and friends. A young lad, somewhat overweight, came up to inquire about this type of work. He got stuck between the shelf and firefinder stand. Since he was unable to make it around the stand and was very nervous about the height, I doubt that he ever applied for the job.

The steps up the tower were steep. I was careful to warn all, as they climbed, about the three headcrackers they would encounter on the way up. ("Don't bump your head on that two-by-six brace." "Duck down or you'll bump your head on the edge of that hole." "Swing over this way so you won't hit your head on the sharp metal corner of the firefinder stand.") It took constant warnings of "Don't bump your head" when several were climbing up, one after another.

I had asked for a remote station when I applied for work with the fire protection association. Lenhart Butte was the only post offered me. I took it, hoping it might lead to something more interesting another year. I had been up only a few days when I received a radio call: "Would you like to be transferred to Lookout Mountain?"

Would I? My heart flipped a flop or two with unbounded joy. Over the radio I answered simply, "10-4." Lookout Mountain has an elevation of 4,500 feet and is in the remote area I had asked for. I knew that if the season was dry I might never be granted days off; nevertheless, I was anxious to make the change. I would leave Lenhart Butte with no regrets. Who wants a lookout station where you listen to the bark of the neighbor's dog, and where things called human beings peer through the windows of the ground house, watching that strange creature inside eating a peanut butter sandwich? My only escape was the top of that 110-foot tower.

I waited one full week, hoping every minute to get that call saying that I was to be transferred. It was the morning of the 13th of July when I overheard a call on the radio saying they were sending a new man up to the lookout, but they did not say which lookout. I waited all day, expecting a call saying they were transferring me; no call came. They must be sending the new man up to Lookout Mountain and I would be stuck on Lenhart Butte all summer. I was sick with disappointment.

It was now 8:30 p.m. To alleviate my disappointment, I called for permission to go down to the Sandy Guard Station, having been told I could go down once in awhile to take a shower or wash my clothes.

While dripping in the shower, I heard Molalla headquarters calling my station. Since I was not in shape to appear in the radio room, the warden's wife, who was dispatcher, took the message and told them I was there at Sandy Guard. It was now well past 9 p.m. The word was that I should be at headquarters in Molalla by 9:30 the next morning, ready to go up to Lookout Mountain.

Needless to say, I dressed quickly and tore off to the lookout as fast as I could. Although it was dark when I arrived, I made two trips up that 110-foot tower to bring most of my things down to the cabin. I then went to bed, but being too excited to sleep, I got up and packed, then went back to bed, slept for two and a half hours, got up again and loaded the station wagon.

It was now the morning after the night of nights. I called headquarters to tell them I was leaving, but was now told to wait until the man arrived to replace me. He did not arrive until after ten o'clock. Since everything was packed in the wagon, I would have to do without breakfast.

The man who was replacing me had just turned eighteen and had never been on a lookout before. There were a few instructions I wanted to give him to help him get started on his new work, such as: how to put the equipment together when it fell apart and the importance of remembering to close the trapdoor. He would have received instructions on how to operate the instruments when everything was working, but only the lookout who has worked with it knows how to read that hydrothermograph when it is out of kilter. Because the tower was so small, with just enough room to walk around the firefinder, that hole in the floor would be a death trap if the door were left open.

After leaving the station, I stopped in Sandy for a combination breakfast and lunch and to buy a few necessary supplies before going on to Molalla where the men were waiting with bated breath. Why had it taken me so long to get there? Well, I might have arrived sooner had I gone without breakfast and lunch, and that after having worked most of the night with little sleep. The fire officer was nice enough, though, saying he had a surprise for me. The ground house on Lookout Mountain was a small trailer. It had a small refrigerator. I could stop in Molalla to buy meat if I wanted to. I told him I must have topographic maps or I would not be of much use up there for awhile. I had not had a chance to buy any for myself. He searched for maps while I went to the store.

The men were at the store, waiting to take me up even before I had the groceries paid for. They jumped into their trucks, I into my wagon, and off we went, through Silverton, past Silver Creek Falls, turning off at the South Burn Guard Station. From there on, the roads grew progressively worse — over rocks, around rocks and through dusty hollows. I had to stay far back from the truck to keep from choking on the dust that rose up in clouds, leaving rocks sharp and bare. Up we went, over House Mountain and down again. Now we really started to climb; it was taking all the power in the lowest gear of my new station wagon. The men stopped occasionally to see if I was going to make it. One switchback was so sharp, I backed up to maneuver around it. Then steeper and steeper it became as we neared the top. The wheels dug in, thumpity-bumping, with no forward movement. The tires were unable to grab a hold on the road until they caught a rock; then with a burst of energy, they made a lurch forward with a gain of a few more feet.

One more switchback and there it was, rising high above the skyline — the Lookout Mountain Tower. I broke out above the trees onto a mountain oasis in full spring bloom — masses of pink rhododendron and creamy white blossoms of bear grass — red dots of Indian paintbrush clinging to a rocky bank and, here and there, patches of many-colored blossoms springing from the rocky soil.

This was the land of my dreams, but this was no time to dream. The car and its contents were layered with dust. I felt like a dust devil, my body benumbed, my mind in a whirl from lack of sleep, hard work and excitement. I was concerned over my new work and overwhelmed by the beauty on this isolated peak.

As the cars pulled up to the trailer door the officer keyed the mike on his radio, "Lookout Mountain, 10-8," announcing that we had arrived and that I was

The Lookout Mountain Tower had large sliding windows that could be opened.

now on duty. It was then four p.m. and time to turn in the weather report. I do not remember the name of the man whom I was replacing, but I will always remember his saying, "I'll take the weather for you as you have a lot to do and are probably tired." How right he was and how grateful I was for his offer! This man will never know how much that little favor meant to me, for I was tired — and tense from the day's ordeal that had begun before daylight that morning. I was trying not to let my feelings show, but I doubt that I could have taken the weather readings and continued them on someone else's report without error, as this phase of the work was relatively new to me. Many factors are involved in recording the daily observations and careful computing is required to arrive at the build-up and burning index.

The retiring lookout began telling me enthusiastically about the deer and grouse that came around the door, while the fire officer was explaining the operation of the radios and operation of the gas appliances (floor furnace, range and refrigerator). I had ears for both conversations while hurriedly writing letters to a few close friends and relatives, letting them know of my escape from Lenhart Butte and giving them my new address. My hand was shaking and it was hard to hold the pen. I wonder now if they could read my writing when they received it. The officer was standing over me impatiently as I wrote. The moment I stamped the last envelope, he grabbed the letters and disappeared through the trees in a cloud of dust.

I was now alone on Lookout Mountain — a rocky knoll bright with the blossoms of rhododendron and bear grass, tips of tall trees peeking over steep cliffs and, to the west, a trail of dust winding down the mountain — dust from the wheels of the officer's truck spinning the loops back down to his world.

This was to be my home for the season. I greeted each day with enthusiasm. It seemed the ultimate of all I had hoped for. The crisp, cool air of morning gave a sparkle to the color and beauty on this small island in the sky. The sharp peak of Mount Jefferson in the Cascade Range was sharply outlined on the eastern sky. White clouds rolled silently across the hills, cascading into the canyons and, as silently, faded into nowhere as the sun took over the day. The song of birds, the quarrelsome voices of chipmunk and pine squirrel, a coyote's distant call, and the constant chucking of the mother grouse to her chicks seemed all the company I needed. When darkness fell, huge racks of antlers would rise over the peak, silhouetted against the moonlit sky. Cautiously at first, herds of doe and fawn would follow, snooping around the trailer, their hard hoofs tapping on an old lookout shutter I had lying in front of the door. A black bear prowled near the trailer at times. I would often see his footprints in the morning, along with the tracks of raccoon and other small animals.

The ground house, a twelve-foot trailer, was parked near the foot of the twenty-five-foot tower. My bed, a davenport at the end, was very comfortable. I had been informed that the cooking utensils and bedding were furnished. Of course I had my own. I tried putting my bedding on top of the bedding already there, but the smell of ancient cigar smoke was strong enough to wake the dead. I wrapped up the smelly bedding in a plastic sheet and carried it to a storage shed that was down an embankment below the trailer. I could now enjoy the fragrance of the flowers and evergreens and sleep in peace in the trailer. It was a nice little trailer, the best living quarters I have had in this work.

I laughed when I started pulling the utensils out of the cupboards. They were cast-offs from fire crews or who-knows-what. Huge, battered tin kettles, bigger than the top of the stove, several two- or more-gallon coffeepots, several twelve- or more-inch, rusted frying pans. And the dishes! Well, the whole mess looked like the aftermath of an explosion in a Salvation Army receiving depot. I did not need a fourteen-inch frying pan to fry an egg on the three-inch burner. Of course, these were minor problems, so I could laugh, for this was still the happy end of the season.

The lookout cabin was a neat six-by-six-foot structure on top of the steel tower. It had large sliding windows that could be opened, but no overhang or shutters. The afternoon sun was blinding and sometimes hot. Big flying ants were a problem here in the early part of the season, so thick on the glass at times that they interfered with my sighting on the firefinder. I was on the tower from the time it was light enough to see smoke in the morning until it was too dark to see in the evening, about sixteen hours on the longest days. I had an hour off at 7 a.m. and at 12 noon to prepare my meals and take the weather observations, then down again at 4 p.m. to take the complete weather report, and an hour at 5 for my evening meal. I had a radio in the trailer, so I was on duty or on call for twenty-four hours, as is usual in this work. It was cold on the tower when the sun went down. I used gasoline, intended for a lamp, to burn in my one-burner gasoline stove. By cutting ventilating holes in a two-pound coffee can and placing it over the burner for a drum, I had a little heater.

This was not a busy station. There was very little radio work due to the relay system. But this, in itself, was sometimes a problem. The warden at Mehama, the station that serviced this lookout, did not call me on direct signal, but on remote through the relay system, which carried the voice to Goat Mountain, then back over to Linn County's signal mountain with a split-second lag between the stations so I could not read his messages. He sounded like a chorus of drunks. I wanted to tell him that, but it's not what you tell a co-worker whose paycheck is bigger than yours.

A boy came up each Friday with forty gallons of water, more than I could possibly use or had storage for. He dumped out the older water and took the empty cans back. Perhaps they used the water for weight in the back of the truck, as they needed weight for traction on the steep climb. (The extra weight may also have been the reason for the many flat tires that were caused by rock cuts on these water runs.)

During the week, long fingers of rust built out into the water from rust spots on the old cans. The water was red with rust particles when I dipped it out, so I saved enough for drinking when it first came up, before it had time to get rusty.

I had left the low country without a chance to prepare for this more remote area where friends could not reach me. I soon ran out of reading material. My friend, Grace, called the guard station for two weeks in succession, asking when the truck would be going up. She had a box of goodies and reading material she and friend Eleanor were sending up. Grace was told the truck was going up on Saturday, but she arrived at the guard station to find it had gone up the day before.

On the following Friday, she made an early call and was told the truck was about ready to leave. Grace jumped into her little red car and drove full speed the twenty-five miles, arriving at the guard station while the boys were filling the last can of water.

This is one time I was thankful for forty gallons of water. Ten gallons less would have been ten gallons late. My dear friends will never know how much that box and its contents meant to me. Not having enough work to keep one busy and little to divert one's mind was enough to exhaust both mind and body.

I brought enough food to last all summer when I came up, so I made very few requests for supplies. Then there were the huckleberries! That hilltop was covered with them, large, luscious fruit that lasted through August and most of September. During my moments off the tower I would make a sandwich to eat after going back up, in order to spend every moment possible in picking and eating berries. They might have been half of my food supply. Never did anything taste so good.

I was surprised to find no provision for trash disposal here, as neatness was a requirement at the other stations. I found where the lookouts before me had dumped their trash over a bank down into the timber. Since the bear, deer, raccoon and grouse were the only friends I had up here, I did not want to clutter their world with cans and broken glass. I flattened the cans and tried to bury them, but could not dig a hole deep enough in that rocky knoll. The bears and other, smaller animals would dig the cans out and scatter them around.

During the early part of the season, I had two carloads of visitors, friends from Salem who had 4-wheel-drive vehicles. How welcome they were!

Another long-time friend wrote to say he was coming up. I knew he could not make it in his car and wrote to explain this to him. It would be another week before I would have a chance to send the letter down for mailing. One day during that week I heard a strange noise, a deep rumbling down in the hills. The noise increased in volume and seemed to be following the road up the mountain. I stood on the tower staring in unbelief as a familiar old car slowly emerged from the trees around the last switchback. It was blowing a huge, white column of steam, followed closely by a cloud of fawn-colored dust. As it crept up nearer the lookout, I saw the pleasant face of my friend peering out between the clouds. He had made it! He laughed when I handed him the letter saying, "No way can you possibly make it up here in your car; please do not try." He was driving a heavily built older car that was too pokey for city driving, but had more power and clearance than his newer one. However, this was more than its old system could take. It had developed "hyperthermia" and was gasping its last when it dropped in its tracks beneath a bank where rhododendrons bloomed in all their glory.

He had not been up long when the boys came on their weekly trip. Again, I was thankful for that forty gallons of water. That is what the old car needed now. Still in convulsions, it was hissing and emitting bursts of steam like a dying volcano. We drenched its tortured throat with water. With a few more burps and gurgles it quieted down.

My friend left after a short visit. I was then tormented with thoughts of what might happen before he got down off the mountain and back to the world again. Long after I thought he should have arrived, I received a call from the South Burn Guard, saying he had just passed there. How thankful I was for that call, for I had imagined him broken down in country he was unfamiliar with, amid unfriendly elements.

This was a frustrating summer in many ways. As the months wore on, my sense of humor grew thin. A U. S. Forest bordered the district just a few miles to the east. I would see smoke rising above a ridge after a storm and, not having a U. S. Forest radio, would not know whether the fire had been reported. Linn County Fire Protection Association was just a few miles to the south. I had no communication with them, either. The Forest Service had planes out after a storm so I could only hope the fires had been spotted. Watching the skies grow red from the forest fires on U. S. Forest timber, I was vitally interested in what was going on and wondering if they were getting it under control, but, busy as I knew the air must be, it was silent for me.

Wildcat Ridge stretched across the near horizon, blocking my view to the southwest. I would see smoke rising above that ridge and have no way of judging the distance. Since there was no landscape in view beyond the ridge, I was not able to relate the smoke to a landmark. I would call it in and be told what I already supposed it to be, smoke from another valley field burn. However, had I not called it in, the powers below would think I had not seen it.

High Camp lookout called in his azimuth on a smoke I had turned in. I heard no more about it, so I finally called headquarters. The girl in the office said the lines did not cross; in fact, they came nowhere near each other. Fortunately I had had my tape recorder on and made a recording of the transmission from High Camp lookout, so I was able to prove the office girl had shot my azimuth from High Camp and High Camp's azimuth from Lookout Mountain. This way the lines would not cross, but take off in opposing directions. I was careful to mention this to my supervisor when he made his one trip of the season up to my lookout. He said it was possible, but of course they had a recording of it. I mentioned that I had a recording also, and had checked it.

My supervisor brought his father along on this trip. While the supervisor was eating his lunch on the tower, I was down eating my lunch and taking the weather. The father spent this hour walking over the peak, peering over the edge of cliffs, paying special attention to the refuse cliff. Was he looking for hides and feathers? Since I never ordered groceries, was I living on deer and grouse? Perhaps the men should have stayed long enough last spring to watch me unload those boxes of canned meat and other food supplies. Or they might have asked, and I could have told them the deer still had its hide and the grouse still had its feathers.

On a night of the full moon I was awakened by a noise near the trailer door, a thumping, scratching sound. It must be a bear, I thought, as I opened the door a crack to scare him away. It was not a bear, but a large porcupine, chewing away on the "welcome" mat. Reaching my arm through the crack, I grabbed hold of the rubber mat. With no more than my arm in sight, the porcupine had no mind to give up his prize. A tug-of-war ensued with me on the winning end.

He will go away, I thought, as I went back to bed, but I soon heard him again, this time working on the trailer tires. I knew porcupines would chew on the tires of logging equipment left in the forest, so I thought I had better put a stop to this right now. I loaded my .22 rifle and stepped out in the moonlight. It was a new experience for both of us. Evidently, the porcupine had never seen a lookout dressed in a long flannel nightgown, making a rifle charge. He took off for dear life, but stopped on top of a rocky outcrop to assess the situation. Was that tall, white creature for real, or just a moonlight apparition? Too late, he was dead. His heavy body rolled down the bank, wedged tightly between the bank and the shed. It was a bad moment for both of us. I hate guns and hate killing. He was too heavy for me to move from the wedged position, so I tried to bury him where he fell, but had trouble finding enough dirt to cover him. I finally scraped up a little gravel from the driveway of the shed and at least got him out of sight. It had been nearly fifty years since I had used a gun, but, after examining the tires on my car and finding that he had also been working on them, I realized I had done what I had to do. The heavy-duty tires were now badly damaged.

A few weeks later, on a dark, moonless night, I heard a porcupine beneath the trailer again. Armed with flashlight and rifle, I sneaked out into the darkness. The porcupine evidently heard my movements and slipped away, for I could see nothing under the trailer, but, remembering the tires, I groped my way down a rocky trail to the car. A feeble shot of light from the overworked flashlight seemed to make the night just that much blacker. Grass had grown up around the car, as it had not been moved since coming up, nearly two months ago. I had it up on blocks, as the tires were soft and there was no way to inflate them here. My flashlight lit up the beady eyes of a porcupine beneath the car, snuggled against a rear tire. While holding my flashlight on the eyes of the porcupine, I stooped down low to see what was going on. I was now sighting the gun upside-down through tall spears of grass, while holding the tail of my nightgown up with my teeth to keep it out from under my feet. It must have been a sight to behold! Dark nights and high mountain peaks keep their secrets well! In a moment it was over, and I had missed the tire. I do not feel proud of myself for having killed these two gentle creatures. I can only console myself by the fact that they are damaging to the timber as well as to the rubber tires. Again, I did what I felt I had to do.

The season wore on and blossoms faded. Smoke from the forest fires mixed with smoke from many field burns down in the valley made for a gray and colorless world. The mother grouse and her family would leave now and then to be gone for several days. How I missed their friendly calls! Always when she returned there were fewer chicks. She started the summer with thirteen and, for all her care and attention to the little brood, the season ended with only three. Coyotes? Bears? Buzzards? There are many perils for the weak on this lonely mountain.

Day after day a tired, burnt-out sun arose over the murky horizon, dragged across a dismal sky and dissolved in the everlasting haze, sometimes giving the western sky a subtle wash of colors, while the planet earth remained a gray monotony. I knew the sun had fallen below the horizon when my feet began to get cold, but I stayed on the tower until dark. One by one the days dropped from the calendar. The buzzards circled slowly, checking the lone figure with an air of anticipation. Two months went by with very little human contact. There was no change in the days as I gazed out over the hazy world below, no line of demarcation between earth and sky. I was beginning to feel the strain. "Am I still on the planet earth, or on some strange sphere where the days have no ending?" Down below one is not conscious of the passing time until jolted by some significant date, birth, marriage or death, perhaps. Up here, I waited impatiently for the end of each day and could almost hear the day fall as it dropped into the past. I reached out eagerly for the next day that I might put it behind me.

It was two months into the season when a call came over the radio that I was to have a day off. A boy was sent up to take me down to the Mehama Guard Station. As the tires on my car were soft, I could not use it.

It was pitch dark when we arrived at the station where my two friends were waiting to take me to Salem. Here, in the city, cars were everywhere and people were rushing madly about, paying no attention to this unearthly creature just arrived from another world. I felt nervous and ill at ease in this strange, busy atmosphere. However, with the thoughtful attention of my friends and a good restaurant dinner, I soon came back to reality. It was one big day off. Then, back to the mountain again, bringing with me a hand pump to apply to the tires.

From my Journal: This was one of the driest seasons on record, but now, well into September, the rain gauge is collecting water and the burning index is down. For me, this meant another day off. A boy came for me again and off we went, slippery-sliding down the mountain. I was called to go back on the following day and returned on a bright, sunny morning to a sparkling, rain-washed world.

. . . The hills have now lost their tired look and the sky is blue again. Time now for the autumn color show. The huckleberry bushes are red, blue-red. The hills are ablaze with their color. They provided well through the long, dry summer. Where the rhododendron bloomed so profusely in the spring, now the red and orange berries of the mountain ash brighten the autumn scene. The vinc maple, unnoticed in its dusty dress of summer, flares out in crimson glory amid the many shades of green and brown. The squirrels are working furiously, harvesting the last, remaining cones. They drop six or eight, then scurry down the tree trunks to bury them in a secret place. The buzzards still circle the tower, but it looks like a summer's wing-work wasted. The old girl is going to make it yet.

. . . It is late September: Rains drenched the forest last night, eliminating fire danger. I have everything packed, knowing it will be the end of the season for me. When I awoke this morning on the last day of September, black clouds were sweeping across the mountains; the canyons were roiling seas of dark fog. At the first light of day I started loading the wagon. It was not raining, exactly. I just seemed to be enveloped in a cloud of water and the wind did not know which way to blow. Everything was mud. There was no way I could keep it out of the trailer as I tramped back and forth in the loading.

I refurnished the trailer with the original pots, pans and bedding. When I unwrapped the bedding that had been stored in the shed, a cloud of blowflies dashed out of hiding and into the storm. Where on earth had they come from? Was there something in that bedding when I wrapped it up last spring? Too late now for small worries; the excitement of going down is on.

As the sun rose higher, a single shaft of light pierced the clouds, lighting a spot on a canyon wall. Then, suddenly, the whole world around me lit up in flaming color. Frantic clouds of orange-red tumbled in confusion over hills and canyons. I was no more than a particle of the elements in the midst of the flurry of sun and clouds in battle. A spectacular farewell to a lonely summer! Soon the color faded and I was "going down." I groped my way down the mountain through heavy clouds that obscured the road.

. . . I am back to earth again with my summer's pay unspent and a heart full of memories. So much is experienced and so little expressed in these simple words, "going up" and "going down." Going up, the sky opens to a season of work, excitement, beauty, frustrations and often days or weeks where space is everywhere, but where you stand in your small crystal tower and time seems forever. You live a summer with no one to bounce your laughter on, no shoulder on which to cry your tears, no modifying influence of thought interchange with other people. Loving the native mountain creatures, you must bear their mistrust of you because you are of the human race. One surely hits the highs and lows in this work, like a kaleidoscope, always turning from one light of life to another. "Going down!" It's a day fraught with excitement. An early fall storm always dictates the end of the season, when nature closes the mountain. There is a sense of release — a feeling of unleashed, unrestricted freedom. I will be with other people again, trying to present a normal appearance to the world, while deep within there remains a silent aloneness.

Chapter 3

Pelican Butte – 1968

With three years of experience to my credit, I was now offered a position with the U. S. National Forest Service. Pelican Butte, with an elevation of 8,036 feet, was surely the top of my dreams. Pelican Butte is in the Klamath District of the Winema Forest in southern Oregon. Lying on the eastern slope of the Cascade Range, it rises abruptly from Upper Klamath Basin with its two lakes, and blocks of farmland reclaimed from the surrounding marshes.

The butte was once a cinder cone; the north and east walls are now completely eroded away, leaving only a cirque-like depression, remnant of the cone that was. A lookout tower rests on the highest point of the now-remaining part of the crater wall. From its base, the wall slopes sharply down to a small mountain lake that lies nestled in what was probably the vent of the old crater.

I drove to Klamath Falls the day before going on duty and spent the night at the home of the head dispatcher, being coached on some of my duties and, no doubt, being scrutinized as to my potential for the work. I knew I was entering a new phase in my life and hoped my past experience would get me "over the hump" until I learned the new work.

At district headquarters the first morning, there was much confusion, or so it seemed to me. Men were milling about, waiting for, or receiving orders. Many of them, like myself, were new on the job. The parking area was full of trucks, ready for dispatch. It was the "opening of the season."

From my Journal: I find the National Forest Service is more diversified, with more activity than the state organizations I have known, and the work, the country and the people all seem very different. The office girls are pleasant and I feel I will like my supervisors: Lynn, the district fire control officer, Jon, the head dispatcher, and Roger, the fire control officer who works out of the Pelican Guard Station. I was pleased to see the good rapport between these supervisors and their personnel, for it engenders good worker relationships and encourages better work performance.

After receiving a few instructions I drove the twenty-five miles to the Pelican Guard station where I will leave my car until the winter snow melts from the mountain road. Here I moved supplies enough to last for several weeks from my car to one of the trucks, while the boys loaded the lookout instruments. Teddy, the packer, loaded his horses onto a truck. Except for a few low places, the road to the snow line was dry. There was only one drift to keep us from reaching the top since snow was light this past winter. We transferred to horses on the down-side of the drift. Teddy rode on up with the pack horses while the boys and I walked. We had only three-quarters of a mile to go and only the one big drift to cross. Above snow line the road was dry and dusty.

We were now at timberline where trees are sparse. Their sculptured forms, bent and twisted, bear evidence of their struggle for existence. The tower, a rough, sturdy structure of man's design, looms from its point on the old crater wall. The boys began their work of opening the lookout while Teddy went back to the trucks for a pack load of water. The shutters on this tower swing up on hinges and are bolted to heavy braces above, reaching over the catwalk. At the close of the season the shutters are lowered down over the windows and locked in place with two-by-fours. As soon as they were removed, Roger crawled under a shutter, opened the door and entered the cabin. I followed closely, crouching low to get under the shutter.

The cabin was dark and gloomy in contrast to the bright, high-altitude sunlight outside. As I entered, I was horrified by the sound that greeted me in the semi-darkness of the cluttered room. Flies! By the million! The noise of their wings was awful. As the shutters were raised I could see them everywhere. The windows were black with them. Everything was full of or covered with flies, dead ones, live ones and all-the-way-in-between-ones, creeping and crawling over each other. They were buzzing and spinning around on their backs in the battered old pans and dishes lying about the room. They covered the mattress, a sad-looking, damp, water-stained object. Flies tangled in my hair and scrunched under my feet as I walked across the room. Musty old girly magazines littered the bed and floor.

Something deep within me wanted to cry and run home, but I quickly subdued that feeling and was careful not to let the men know that anything bothered me. I went ahead, preparing for my work, taking it all in stride. I was glad I had disregarded that part of Jon's letter saying "You will not have to bring dishes or cooking utensils, as they are provided." I asked the men to put the old pans and dishes in the attic and dispose of the girly magazines. Roger teased me about not wanting to save them. Well, for the time being, at least, I could put the foodstuff in the oven where the flies could not get in. Oh, no, I couldn't. The flies had thought of that, too. They could not get in, but they surely died trying. Their smashed and dried carcasses stuccoed the inside edge of the door.

By the end of the day I had things cleared up a little and had spread my plastic bedroll cover between the bedding and mattress. I was tired, but happy and excited about my new work. Everyone I've met so far seems pleasant and cooperative. I had no illusions of an easy job when I signed on here, and easy it is not.

This is a busy station. There are many sources of fire: summer homes, mountain trails, campgrounds, logging operations, a railroad, five open-pit garbage dumps, and many electrical storms that often keep me up at night. I have five radios to monitor. Even though the messages are not all for me, I have to remember and make notes of many, as this is the central point for relaying messages to the workmen who cannot be reached from the district office. I must know the location of the workmen, trail crews, and patrolmen, as well as all observation planes that may fly over for I never know when I might be called upon for this information.

I find I have much to learn and, what is more confusing, I have much to unlearn. The radio procedures and use of code numbers differ from what I have known on other stations, making my previous experience more of a hindrance than a help. I have been called down on the air a few times for improper methods. This was not by my supervisors, Lynn or Jon; somehow they seem to understand. It came from a few in the lesser echelons whose minds are imprisoned within this one district where they have worked for years.

One of the fellows, Bill, is particularly caustic in his orders. I have heard that he was once an army officer and is used to bellowing at his subjects. I record him as "Billy Goat Gruff" on the radio log. Many of the seasonal workers come under his aerial blasts. Whenever I am called to relay a message to one of the boys, I strain out the "ugly" before relaying it on. I have no way of knowing who hears the original message, but was complimented by one of the boys on my title for Bill on the radio log.

Being unfamiliar with the country, I spend much time in studying maps, relating them to the surrounding country, learning the names of the creeks, mountains, ridges, etc., in a fifty-mile radius. I must correctly relate each ridge and draw to the topographic map. The firefinder is used only to locate the angular degrees from this point; I rely on topographic maps to judge the distance and get the legal description. I must know the location of trails, campgrounds, lakes, logging operations and other areas where fires are likely to start, then use this knowledge in making my judgment before turning in a smoke (and hope I am not far off.) Not all smoke is a fire to be reported; if I am not sure of its source, I turn it in and a man is sent out to check it. I may see smoke rise over a ridge. Is it behind the first ridge or on one of the many beyond, not visible from this station, but traced on the map in contour lines?

Smoke from a mill or other legal operation may drift along the ground, then rise from behind a ridge to become visible some distance from the legal description which I have on record. Smoke may rise on the azimuth of a known campground and be putting up more smoke than a campfire normally would. Is it, or is it not out of control? I may spot smoke near a house or on a farm. Has a fire started up, or is someone burning without a permit? If it is an illegal fire, it will probably be out before the boys arrive to check it, so, "what does that lookout up there think she's seeing?" A dust column seen through blue haze sometimes has the appearance of smoke. I study its character and movement before making a decision. I am the only lookout on the Klamath District; there are a few rules that apply only to me. I must have a full report before turning in a smoke: the approximate distance, legal description, local landmarks, character of the smoke, access roads if any, as well as the azimuth. The other districts allow their lookouts to turn in the azimuth first, then go off the air while they look up other information. This gives them a better start in reporting a fire, but Lynn is a stickler for his method of operation, thinking it will get the crews out faster. He has a good crew and expects the best of them.

I found an error on the firefinder map where a section had been misnumbered. It was in an area where adjustment is made for curvature of the earth. This was a serious mistake as it put the land description one mile off when I reported a smoke. I discarded this map after Lynn had studied it, and used the only other one available. This was an old map that I found mounted on the firefinder beneath the one I discarded. It was completely blank in the area west of this station. I filled in the missing area with a section cut from a tourist map and Scotch-taped it over the blank space. Since paper tends to expand and contract with changes in humidity, it is impossible to get the exact legal description when reporting smoke. (I was to work with this map for the three years I spent at this station.)

There is also misinformation on the two road and trail maps, making it necessary to compare maps before turning in a fire.

If Lynn is not able to get better equipment, at least he is understanding. He has a sense of humor, often directed at himself. No matter what excitement or predicaments arise, he is always pleasant and considerate of those working under him.

. . . It is two weeks into the season now. The big drift has melted out enough to let cars through. Roger took me down last evening and I brought my own car back. It became dark before I reached the lookout. On the last few miles of road there were rocks and muddy ruts to dodge and sharp switchbacks to feel my way around, switchbacks that touched the very edge of the crater wall. The wind was blowing a gale when I arrived on top. It whipped through the old frame of the tower as I fought my way up the steps while carrying supplies from the car. Dropping the trapdoor down behind me, I entered the darkened cabin, listened to the rage of the wind and felt its force against the tower. What a fortress the tower seemed.

. . . A few evenings later: I went down for my walk after having signed off for an hour. An atmospheric haze was moving in from the southwest. A sudden blast of wind fairly blew me back up the mountain. I was enveloped in a gale of lava sand and gravel. Spicules of ice now added to the gale, slashing everything in its path. The ragged trees, unyielding, appeared as ghosts, their naked branches reaching out into the fading light of the storm.

I thought of the car: "Will it freeze up?" I jumped in quickly and started the motor. The red temperature light came on; the motor was hot and the water not circulating. I rushed up the tower steps, grabbed a blanket off the bed and a tarp, while the storm raged in fury. The wind blowing against the car was all that enabled me to get the blanket and tarp over the radiator and lashed to the wheels with twine. I started the motor again and in short time it had thawed out. The car was now coated with ice and lava sand. I called for permission to take the car down for antifreeze. Although I had some with me, I would not try to add it in this storm, as it would mean reaching far under the car to drain out some of the water.

Unable to see through the windshield, I kept the car on the road by sticking my head through an open window until I got down into the trees where the ice and sand soon melted off. I spent the night at the Pelican Guard Station, then returned to the tower this morning.

It must have been wild on the mountain last night; the tower was encased in an armor of ice and two or three inches of ice had formed in the cans of water. The sun is now shining brightly. With a fire going in the heater, it did not take long for the ice to soften enough for me to scrape off the windows, using a cookie sheet for a scraper.

Well, the storm helped in one way: it discouraged the flies. They disappear when the temperature drops below sixty degrees or when the wind is blowing hard.

. . . Two of the boys on the trail crew, unfamiliar with the country, took a wrong branch when hiking down to meet the truck that was waiting to bring them out. It was too late in the day to send searchers, but men were posted at every trailhead where there was any possibility of their coming out. I received some lively messages from Bill to relay. The fellow waiting at Cold Springs, five miles down from the tower, called a few times asking, in effect, "How long do they want me to stay here?" It was dark, well past mealtime and getting cold, but Bill's replies were heating up considerably. I changed the tone of his messages before relaying in simple words, "You should stay until notified to leave."

"Alice, what did you have to eat for supper?" It was the voice of the man waiting at Cold Springs trailhead.

Before I had time to think, I answered the truth: "V-8 juice." I had been too tense to prepare and eat a meal. I immediately realized the gist of his message. That human being behind the mike was getting tired of sitting in the dark and cold with nothing to eat. (I never did meet the man behind the mike.) I called back, saying, "Come up and I'll fix you something to eat." Now, I might have found myself in trouble, for it is against regulations here to send out personal messages. A lookout should not hold personal conversations with, or give orders to other workers without permission.* Of course I knew he could not come up anymore than I could leave my station to take a meal down to him, much as I would have liked to. But the message did break something loose down at the district office. Jon's voice came on, for relay, saying they would send someone up to replace him. It was the first pleasant voice I heard through this ordeal.

Later, about one o'clock in the morning, Bill called. "Alice, you might as well go to bed and get some sleep. There is no use spending more time on those fellows tonight." I was shocked to tears at the gentleness in his voice as I stood staring over the cold, dark, midnight hills, wondering where those boys could be. (They came out the next morning, across the mountains to a ranger station on the Rogue River Forest.)

August 14th. . . The morning weather report predicted rain. Later I received a call from Lynn: "In view of the weather report, you can take your two days off. Do you want to come down tonight?"

"10-4," I replied. (Oh, happy day, down tonight, 10-4 by all means!) Then, "I'll leave after I sign off this evening." That was a day of anticipation. Roger called, saying he would meet me at the lookout road-head with my paycheck and mail, saving me the time of going back to the guard station. This way, I could head out for the freeway and be home in Salem by midnight.

As I made last-minute preparations for going down, I felt an urge to get out of my lookout togs. Crumpled away in the back of the under-bed drawer was a pink flowered dress, never worn. Since I was going down to earth and the weather was warm, why not dress before I went down? A pretty dress does much for one's morale, although I must say I felt a little out of place when I rested the heavy trapdoor on my back, lowering it as I backed down the

* There is good reason for this; the channels must be left open for business and emergency calls.

steps. I tossed my backpack into the car and took off, feeling somewhat like Cinderella in her pumpkin coach.

Down the road a bit I came upon a game warden, parked at a switchback on the edge of the crater wall, watching for poachers, I suppose. He gave me a questioning look, but since I did not look like a poacher, I was going to pass him by, until he made a move to get out of his pickup. I then stepped out and explained my presence. I was the lookout "going down." (Don't ever try to stop a lookout "going down.")

Further on I came to a section of road in the worst stage of reconstruction. It was being rebuilt by a lumber company into a timber sale in preparation for logging. There was no road at this time, just a mess of loose dirt and roots. My car was hooking roots and dragging dirt up to the bumpers. No matter — I was going down.

I was not completely out of this when I came upon two big trucks, dead still, side by side, blocking the would-be road. The occupants seemed to be having an argument. I waited a few seconds, thinking one or the other would back out of my way.

As I watched the two stone faces, I seemed to read their minds and imagined their saying, "What's that old lady doing up here? Let her get down if she can."

Well, I knew I would get down, but I did not know just how. I would try the easy way first. Getting out of the car, I waded through dirt, ankle deep, fully aware of my ridiculous appearance; this was no pink-dress occasion. I stepped between the towering trucks. Looking up at the nearest driver, I said, wistfully, "Will you please move your truck so I can get through? I am the Pelican lookout. A forestry man is waiting for me down on the highway and I don't want to keep him waiting."

A continuation of the stony glare seemed to say, "So what?"

At that moment a thumb popped through the open window of the other truck, making a downhill arc, directed at the driver I had spoken to. The thumb must have had clout, for immediately the truck backed, leaving room for me to pass. No words were spoken by the men. All was in pantomime. Perhaps my referral to the Forest Service had the open-sesame effect.

Now, with that obstacle behind me, I tore off to the highway in a cloud of high spirits and dust, picked up the envelope from Roger and set my wheels spinning for Salem. I might have felt like Cinderella, but those princes back there in the trucks did not have my glass slipper, of that I was sure.

I spent my two days at home. Then, leaving Salem at 2:30 on the morning of August 17th, I came back to the tower. I ran into heavy fog in the mountains, so I took the freeway to Medford in order to miss the heavy trucks that would be hard to pass at night on the single-lane road to Klamath Falls. I became drowsy and actually fell asleep for a fleeting second, nearly driving off the road. I stopped at a campsite, slept for ten minutes, then drove on, refreshed, arriving at the Guard Station by 8 a.m. Roger brought me up to the lookout for Jon did not think I could get my car up over that road.

A storm came up in the afternoon, rain with a driving wind blowing water in around the windows. It ran under the bed, then flowed across the room. Toward evening I began receiving calls from the patrolmen and trail crews who were out in the Mountain Lakes and Sky Lakes area, all asking permission to come out. No prediction of a storm of this severity had been made when they went into the mountains earlier in the week, and they were not dressed for this. The storm was fierce and getting colder. The driving rain, now crystals of ice, was sealing the windows, stopping the flow of water into the cabin. The boys in the Sky Lakes area hiked miles down to their rig at Cold Springs only to find that someone had stolen their gas. It was a miserable night for them; their clothes were wet and freezing. I had many messages to relay while the storm slashed at the tower and rattled the shutters so loudly it was difficult at times to read the messages. The boys were instructed to use what gas might be left in the tank and coast wherever possible. No one seemed anxious to go to their aid. "See if you can make it down."

I waited up until 11 o'clock, expecting to hear or receive a message saying they had made it (or had not), but no call came. Finally, turning my pillow dry side up, I dropped into bed, tired from my day that had begun at 2 o'clock that morning. Sleep did not come immediately. Entranced by the rage of the storm, I lay awake, wrapped in darkness — no light but a faint glow from the heater, turned low. For hours I had listened to the rattle and bang of the shutters. Now, with the radios quiet and my pillow against the windowpane, I listened drowsily to the undertones of the storm: the clinking of minute slivers of ice, like a soft tinkling of tiny chimes and the whispery sound of icy crystals brushing against the glass. The storm loosened a wing-nut on one of the shutters. How it shuddered and squawked as it tore at the hinges and banged on the brace. The sturdy old tower jerked and swayed just a little with the gusts. Well, let it blow; I felt cozy and safe inside, only inches from this wicked storm, but where were those boys? Did they make it down? Why didn't they call?*

A few hours later I awoke with a start from a deep sleep, landing on the floor in my wool-socked feet into a roomful of black, acrid smoke, a red flame shooting up several feet from behind the heater. The room was like a furnace. I was wet with perspiration and choking on smoke. Outside, an ice blizzard raged through the night.

My only thought: "Get down and turn that gas off at the tanks." I threw on a warm coat and sheepskin slippers and dashed to the door. It was sealed with ice — one kick broke the ice, allowing the door to open. Everything on the catwalk was sheathed in ice. Needles of ice stung my face and hands as I fumbled in the dark for the lifting

* I was told the following season that the boys did make it down, near midnight, to a closed gas station near the lake, where they awakened the attendant to get gas that would take them back to their Lake of the Woods Station. Wet and cold, they were met by the usually friendly attendant with an unfriendly shotgun. The attendant, thinking it was a stickup, was set for the occasion. However, the boys were able to identify themselves and did get their gas.

ring on the trapdoor, but the door was frozen into the catwalk beneath an inch or more of ice. I was trapped between fire and ice. It would only be minutes before the heat would break a window. I felt my responsibility to the tower as I envisioned it being licked to the ground by the gas-fed flames whipped by the devilish wind, but I hoped the fire would melt the ice outside and let me down ahead of the flames. Now I could only call for help. Dashing through smoke and blobs of soot, I reached the radio. "The cabin's on fire and the door's frozen shut," I shouted, losing my cool.

The men must have hit the floor as quickly as I had for they left in their rigs immediately. Roger, from the Pelican Station, had seventeen miles to go up the rough mountain road; Lynn, from his home in Klamath Falls, had forty-two miles, some of it sharp curves around the lake, posted at 25 miles per hour, which I am sure he did not observe. Roger, now well on his way, called on the mobile unit, "Use your fire extinguisher."

"I can't — I can't get near the base of the flame."

"Use it anyway," came his reply above the noise of the truck. I knew the directions on the fire extinguisher read, "Direct the extinguisher at the base of the flame." There was no way to reach the base of this flame because the heat was intense and the base down low, behind or inside the heater. I directed the extinguisher above the stove and was surprised when the flames went out. Raw gas was now gushing up through the heater, while two pilot lights flared uncontrolled on the range and refrigerator.

Now in the total blackness of night the sounds of the storm were frightening: the roar of the wind and the waves of icy crystals dashing against the tower cabin, the desperate thrashing of wind-blown shutters and the threatening hiss of raw gas gushing up through the heater. I now felt real fear for myself. A flick of time on the wrong side of a second and the tower would blow into the night. I knew I had to move or I would be moved, in pieces, so I spared no seconds in getting those pilot lights out before they ignited the gas. Only then did I think of the claw hammer that I use to pound the lids off water cans. I grabbed it from the windowsill where it lay and rushed out to make another attack on that trapdoor. What with the wind, the darkness, and the slippery ice, it was a fight to keep on my feet, but I managed somehow to break enough ice around the edge of the door to get the long claws under and raise it up. There was no way to fasten it to the wall as the wall was crusted in ice and the catch embedded within it. I dared not let the door down behind me, knowing it would freeze over and I would be left out in the storm. Jerking a ten-gallon water can loose from the frozen wall, I laid it on its side and rested the door on that as I slithered down the icy steps. After closing the valves on the gas tanks, I clawed my way back up the steps with nothing but ice to hang onto. I had lost all feeling in my hands from the cold. Freezing fog had built slabs of ice four to six inches wide on the railing and net-wire surrounding the catwalk and along both sides of the steps.

It was now showing light in the east and the storm was letting up. I re-dressed, putting on clothes I had omitted the first time, then called to the men to let them know everything was under control, and that they need not come so fast they might break an axle, as someone had done a few days before.

Roger answered: "Are you dressed?"

My answer, "Dressed and ready to receive my rescuers."

When Roger arrived, he turned on the valve and touched a match to the oven. In a burst, flames filled the oven cavity. The trouble: a regulator on the tank was frozen, releasing gas, full pressure, uncontrolled, into the cabin appliances.

As daylight broke, the wind died down and the clouds settled into the valleys. The sun came up in a clear, cold sky above the sea of clouds where only the tips of the higher peaks were showing above the turmoil. When Lynn arrived, breaking out above the clouds, the peak was a-shimmer in frigid sunshine and sparkling ice. I must have looked like a chimney sweep.

While the men uncoupled the stove and were carrying it down to the truck, I stepped out with my camera to record the work of the storm. The entire tower structure was encased in a rough coating of ice. Long, jagged spears of ice jutted from railings and from water cans iced to the wall. A long shadow of the lookout tower fell across the earth-bound clouds. Rainbow colors, awash on the mist, encircled the tip of the shadow. The air was cold and quiet as death.

Well, the stove and I both had the day off to go down for repairs and cleaning. Lynn would have given me two days off to get my nerves settled, but I felt anxious to get back to the tower. The road was now packed down enough that I could bring my own car back again.

I feel now that I should have used the fire extinguisher before calling for help on the radio, but Lynn said, "You handled things well: you called us, you got the fire out, you thought to get the pilot lights out, and you turned off the gas; you did everything right." Everyone was so nice that it made me feel better about the near-tragedy. Lynn said he should have had antifreeze in the tanks, but who would expect this to happen? It had never happened before.

August 19th. . . The storm took another swing at the mountain last night, leaving several inches of snow and temperatures in the low twenties. This was a freak summer storm that also struck the lower elevations, closing a few of the mountain passes. When I awoke this morning, the valleys were cloud-filled to mountain-top level. Tumultuous clouds raced overhead, trailing ragged veils of mist. As the rising sun shot its rays above the horizon, the moisture-laden air burst out in golden light. Dressed for the cold, I went down from the tower into an aura of dazzling light. Golden veils of mist were rising to the warmth of dawn above the cloud-filled cirque, with higher clouds sailing overhead. I felt suspended in a nebula of golden light, beyond the realm of man. No words pertaining to the earthly realm could ever portray the magnificent power of a storm-swept mountain sunrise:

beauty, almost more than one can bear. In a fleeting moment of a single dawn comes memory for a lifetime.

August 20th . . . The clouds have been spectacular today, surging over the peak in a frenzy, only to dissipate beyond the mountain. There is no fire-watching today for the storm dampened the forest and erratic clouds now obscure my view, but I have other things to do. The fire left a film of oily soot on everything; cleaning this up will take several days. There is no shortage of water since I have only to step out on the catwalk and flex the netwire to fill my large pan with the slabs of ice, then bring it in to melt on the heater. The ice is so solidly formed on the wooden part of the rail that it cannot be broken off, not even by use of the hammer. Lynn sent up a sack of salt pellets to use on the steps, but this did not work; the wind whipped them off to the sky before they had time to melt in. I tried dissolving the salt and pouring the water onto the steps. This worked fine as far as the ice was concerned, but later I looked down to see my black car white with salt crystals where the wind had blown the saltwater over it. I went off the air: "10-7, to wash the salt off my car."

Sept. 3. . . Cumulus clouds were building all day. By late afternoon a huge anvil had formed, covering most of the eastern sky. As the sun went down, it reflected its light to the eastern clouds, gradually tinting the stark white anvil a bright cerise. Beneath the anvil, a dark blue haze obliterated the landscape. The stage was set for a lightning storm. Slowly the color faded into the pitch-black night; streaks of lightning split the darkness as the storm passed over and beyond the Winema Forest. I remained on duty well past midnight, watching the course of the storm, recording as accurately as possible any strikes that hit the district.

Sept. 7. . . Around 5 p.m. yesterday I spotted a column of smoke rising above the treetops. It was only a few miles below the tower in an area of no known activity. By its character, location, and because of the recent storm, I would have judged it to be a sleeper come to life, but I knew there had been no strikes in this area.

Roger made a flying trip up to check it. He reported it to be smoke from a logging company's road equipment. The company had started building their road into this timber sale without prior notice to anyone. At about the same time tonight a column of smoke rose again in the same spot. I reported it, saying it was probably the same equipment, but it looked too much like a smoke to go unreported. I then snapped a picture for evidence. Roger was unhappy at having to chase up there again, since he was just ready to go off duty for the day. He had me watch the down-road for dust so he could make the curves on the fly without fear of running into a logging truck or Cold Springs tourist. One may follow a vehicle on these mountain roads by the trail of dust rising above the treetops. When two vehicles meet, someone has to give. Of course, this smoke was reported again as equipment.

(I later sent a picture to Lynn, who in turn showed it to Roger. He stared at it in disbelief. "Well, I'll be — I wouldn't believe it; that sure looks like a fire." Now, for once in their lives, these earthlings had a lookout's-eye view of a false smoke. I made a note on the back of the picture: "Pity the lookout who cannot tell a forest fire from a caterpillar tractor." I still wonder, though, if that cat operator wasn't frying bacon down there and if, after seeing the picture, Roger did not think so, too. The smoke was gone long before he arrived to check it, so he had only the operator's word for the source of the smoke.)

We have had many electrical storms during the past few weeks, setting off a number of fires. Many nights I have remained on watch as the storms slowly strike their way to the north. Usually building up in California, they travel in a northerly direction over southern and central Oregon. My day starts again at the first light of dawn. I am ever alert for the wisps of smoke that might arise above the treetops or over a hazy ridge. There is always an air of excitement and tension when a storm is building up. If it strikes during the day, the district office makes frequent calls regarding the progress of the storm. "How does it look from up there? Which way does it seem to be moving? Does it look as if it will hit the district tonight?" The office and my station must keep in close contact with the fire crew. Silence reigns while the storm is passing over, then all radios come alive. The airwaves jam with fire calls as little puffs of smoke arise above the treetops here and there. When a tree is struck by lightning it may immediately set the forest afire, or it may send up a puff of smoke, then die down and not be seen again. It may smolder in the forest duff, then fire up several days later. Sleepers usually come to life in late afternoon when humidity is at the lowest point.

. . .I turned in a smoke on Squaw Lake one evening recently. It was along the Skyline Trail. "Too much smoke for a campfire, but not spreading," I reported. Roger hiked in several miles to check it, only to find it was a large bonfire for a group of twenty-four campers. I thought, now this will really make Roger mad. Later, he remarked that he likes to walk alone in the forest at night. I now realize this happy-go-lucky, sometimes crabby Smokey the Bear has deeper thoughts that do not surface in his workaday world. How little we know of the folks we meet and work with day by day.

. . . How does one record the endless days of cloudless skies and hazy sunshine, the "warmer and drier" days that may come toward the end of summer. Twice daily the forecast comes over the teletype to the supervisory office and is broadcast over the forest radios, always ending with the words "warmer and drier." These are "red flag days" when fire danger is extremely high. There is a tenseness you can feel; you can sense it in the voices on the radios. I am watching constantly during the hours of light and have formed a habit of waking up for night-time checks. The fire crews are on standby or must leave word where they can be reached if needed. The sky is blue-gray and the earth is gray-blue, making ridges hard to define and smokes hard to pinpoint.

Grass fires sometimes start along the railroad right-of-way, ignited by the hotbox of a train. These fires have always been spotted by ground crews or tourists traveling highway 97. They have caused me plenty of trouble as I cannot distinguish the thin blue-gray smoke through the heavy pall of blue-gray haze that fills the lower atmosphere. A few days of this and tension builds up within me. I feel tired and crabby, but have no one to crab to.

Five o'clock brings a little relief, for this is the time when Lynn calls up his work orders for relay to the boys in the field and those at the three guard stations. If he has had a bad day and is crabby himself, it sometimes brings out a bit of his humor. It is about this time of day, also, that the boys in the field call in their orders or requests. The orders, which are taken in by pack horse, might be anything from mousetraps to dynamite, their requests usually, "will we have our days off?" It is a release of tensions to exchange conversation with other human beings, even if only on business.

After these calls are completed, I usually fix myself a bite to eat. Then, around seven or eight o'clock, if there are no fires or other emergencies, I leave the tower for an hour or so. I may go rock hopping down to a natural rock shelter at a point below the tower, or I visit the tree characters on the peak. Last night there was a change in routine.

I heard a car grinding up the road. As it came to a stop at the foot of the tower, the doors burst open, emitting three wild kids and a hound dog. Screaming and barking, they scrambled up the steps, followed by Pa and Ma. I was crabby enough to make the stay short for the dog that barked excitedly at the voices on the radios, but I did not have much luck with the kids. The ruckus caused me to miss a call from one of the boys who had hiked in to check a smoke I had turned in.

Pa was thrilled with the lookout business and asked how he could get his wife a job as a lookout. "The kids would love it; they'd have a ball up here. I could come up on weekends and fly my model airplanes from the catwalk," he said with great enthusiasm. The kids were already having a ball. While one was seeing how far he could lean over the back of the stairwell before his mother could make him stop, the others were running up and down the steps, bringing up rocks to throw down over the crater wall. I did not give him much encouragement in applying for a job, not at Pelican, at least. I could not picture Lynn hiring this wild mob.

It was only the evening before, at dusk, after I had had my walk and was tired and ready for bed, when a car drove up. Looking down, I saw a family with several children. Excited and happy, they fought with the wind that blew sand in their faces. I could not see them disappointed, so I dressed, opened the door and asked them up. I was glad I did when I found it was Tom and his family. Tom was on the recreation crew. I had known him previously only as a voice on the radio.

Sept. 12. . . The seasonal workers have been leaving the district in recent weeks, to return to their schools and colleges. The permanent employees have been shifted to cover their work. This morning, a Volkswagen bug crawled up the mountain, loaded inside and out with the summer gear of a young mountain patrolman. Jerry had spent the summer alone, patrolling trails in the Mountain Lakes Wild Area, camping out wherever his work might take him. Emerging from the pile of baggage, he introduced himself, then added, "I've been talking to you all summer on the radio, so I wanted to meet you before I left."

I was pleased to meet him. I felt concern for the boys all summer out on the trails in all kinds of weather, with mosquitoes galore and mice in their rations. It is not an easy job, but no one looks for a life of ease on a summer job with the Forest Service. Some of the boys called their good-byes on the radio, perhaps adding a note, "I am glad to have met you," or "You were nice to work with." Well, I certainly enjoyed my association with them, whether I met them in person or only as voices on the radio. They are nice kids; I hope life holds the best for them.

October 6, 1968. . . I got up to make my near-midnight fire check. The night was aglow from a billion stars and a partial moon. Cloud veils were drifting by, silently, like sheeted ghosts in the night, trailing dark shadows across the hills. I turned my flashlight on the clock. Twelve midnight — tomorrow is now today, my sixty-fifth birthday. When I awoke again, light was showing in the east, the mood of night soon washed away by the light of day. Fog settled in, visibility zero. A good day for writing letters, washing clothes or baking bread. There would be no firewatching until this burned off.

Later in the day Lynn called, saying, "I notice Pelican Butte has had a little skim of fog over the very tip all day. All the other mountains are in the clear."

"What have I done to deserve this? What a way to celebrate my sixty-fifth birthday."

"Oh, is this the day? Well, Happy Birthday anyway." Then, with his voice trailing off in a chuckle, "Everyone else is in the sunshine."

"Well, thanks, at least this is a unique way to celebrate one's sixty-fifth birthday, socked in on the top of an 8,000-foot mountain."

. . . It is now late in the fire season and the nights are cold and frosty. The earth below is washed in bean-soup haze. Smoke from piles of burning slash drifts lazily beneath a temperature inversion. But I still have the sunshine with me here; the morning sky is blue and my realm is white in frosty splendor. I raise the flag at sunrise each morning as though all the world were here to see it.

This morning I pulled on the rope which raised the flag, while trying to dodge the shattered fringe of frosty lace that clattered down around me as the rope slid through the pulley. With my hands freezing, I turned to skitter up the tower steps to the warmth of the tower cabin, but found my steps blocked by a man with a gun. I was startled, to say the least. Intent on my ground work, I had not seen him slip through the trees and over the brow of the peak.

As I stood gazing at the barrel of the shotgun, he asked casually, "Have you seen any grouse around here?"

"No," I replied almost truthfully. (I hope he doesn't see the tracks where a shy little family had crossed the road, one switchback down below.) I decided I had nothing to fear for myself, as he was just a misplaced hunter, but never before had the form of mankind slipped through the silence of this hour in the morning.

There is little wildlife to attract a hunter at this elevation on Pelican Butte, perhaps because there is so little vegetation. A few plants have dared to raise their blossoms above the lava sand and rock to absorb the rays of the summer sun, only to be buried in ice and blasted by wind and sand. They bear little resemblance to others in their genera, but in their small way stand out, each as individuals. One looks at a tree in a lower environment and sees it as part of the forest. Timberline trees, shaped and strengthened by the storms, have developed rugged individuality. We stop and study each one as a separate entity, admire the beauty of its form and character and wonder about the story of its life.

. . . A cold, gusty wind was blowing all night long. Shutters straining at their hinges shuddered from the blasts. A thin cloud cover delayed the early light of dawn. I took this to be a wave or lenticular cloud; except for the deep blue color, it had the appearance of spun glass or "angel's hair." Arched over the peak, it reached across the lakes and marsh. Dark, wispy fingers trailing earthward from the filmy sheet vanished when they reached the lower atmosphere. Light from the sky above penetrated the cloud, tinting the ceiling of my world a deep and stormy blue while permeating the atmosphere with cold blue light. The wind was still blowing an icy gale as I rushed onto the catwalk with my camera, too late to catch the streamers sweeping over the lake, but just in time to witness a violent clash of color as the vibrant rays of a crimson sunrise touched the deep blue veil of the cloud, reflecting its light onto the lakes below. For one exciting moment the bright crimson light skipped with the ripples on the wind-blown waters of Klamath Lake. Drainage canals on the marsh were now ribbons of light. With my hands freezing and tears in my eyes, I stood transfixed until the last trace of color fell to the gray of a cloudy day. The splendor of this dawn will remain a part of me forever.

. . . The clouds have been heavy and rain was on the forecast this morning. I have everything packed, knowing I will soon be down for the season. Many of the lookouts on other forests, as well as from Crater Lake, are down. I have had little radio communication with the other lookouts, so there were no good-byes. I knew they were down when their peaks became silent. Lynn called this morning, "Alice, it looks like this is it; I'll send someone up to bring your things down. Jon and I will come up later to close the lookout."

The predicted rain began pelting. Boxes, packed and ready to go down, are absorbing water as rain floods the floor — little does it matter now. I am going home. The strain of a long summer over, I am free again. This was a hard summer in many ways. It was also a beautiful, exciting and interesting summer in an entirely different atmosphere, the beginning of another chapter in my life.

Long fingers of ice projected from the cat walk railing, the result of a freak August storm.

There was no shortage of water after the ice storm. By flexing the net-wire that surrounded the cat walk the large pan could be filled with slabs of ice that might be 4 to 6 inches in width. The ice would then be melted on the heating stove.

These lacy appearing ruffles of ice were designed by the wind during a night storm. Later, after the wind died down, the ice was coated with a layer of frost. What you see here is frosted ice on the net-wire surrounding the cat walk.

After an ice storm — the lookout shutter was left with a rough crust of ice. During this storm the ice built up to 19 inches on the bolt that supported the shutter. About 10 or 11 inches shown here.

If it's not a lightning fire, then what is it? A catapillar tractor? Or is the cat operator frying bacon down there?

A glow of rainbow colors encircle the tip of the tower shadow where the shadow lies across earth-bound clouds. The colors, awash on the clouds, are in constant motion, changing in form and brilliance with the motion and density of the clouds. This phenomenon is called a "Glory Ring." Ice covered trees stand out here in the foreground.

A bad storm can mean time off for the lookout. With my backpack loaded with two days supplies, I slithered down the icy steps to meet the truck that was waiting below snow line to take me down.

The sun comes out in the wake of a storm revealing timberline trees in a glitter of ice against a bright blue sky — beauty born on the winds of a storm. Down below the moisture falls in the form of rain.

Chapter 4

Pelican Butte – 1969

From my Journal : How quickly the mileposts of life flash by when one is surrounded by life and activity. I am back on the mountain again where the brakes on time are applied. Here on Pelican I do have contact with humanity, although still locked on the mountain, alone within myself, living a life that is hard for others to understand. But there is an invisible force that calls me back to this existence. I would have been disappointed had I not been asked to return.

I came down to the area a day early in order to acquaint myself with a small part of the country the tower overlooks, exploring the campgrounds and guard station at Lake of the Woods and along the lake basin to the Seven Mile Guard Station. How different is perspective when seen from the earthly point of view. I spotted a fire at the county dump and sped to the Ranger Station in Klamath Falls to make a report, since there was no sign of life at the Pelican Station. I spent my first night in a Forest Campground where I removed a few things from the station wagon and hollowed out a space in the back, ample room to tuck myself into, and had a fair night's rest.

After another day of exploring I returned to the guard station, ready to go up in the morning. I unrolled my sleeping bag on the battered front room hide-a-bed where any creature in standing with the Forest Service is welcome to spend the night, whether it be a fawn, a coyote puppy or merely a human being.

I believe Roger's preferred guests are his animal friends; at least he does not complain about the mud they track in. He is in charge of the guard station, although all employees of the Klamath District are privileged to enter at any time. The tiny fawn has no key; however, with its cute little face and winning ways it does not need a key.

The coyote pup is shy and has not adjusted to the unnatural habitat, although his life was saved by the boys when his den was caved in by the wheels of a tractor, killing the other pups.

Roger keeps the cabin clean in spite of the in-and-out traffic of forestry men. It has a touch of man's decor, such as kitchen drapes made from the pull-tabs of beer cans. The cabin consists of a front room which serves as office, radio room, meeting place or whatever the need might be. It contains a wood heater,* the hide-a-bed and a few chairs. The fire officer's room and the kitchen are off the main room. A lean-to shed has been attached to the back and fixtures added on the far off-end to make the cabin "modern," leaving a strip barely wide enough for a cot and a runway from the kitchen door back to the "modern."

Workmen or weary firefighters would toss their sleeping bags on the cot and find rest and peace for the night, or perhaps rest for a piece of the night should a fire break out.

We were all up bright and early next morning. After eating a hearty breakfast that Roger had fixed, we prepared for the assault on Pelican Butte. There was more snow this year; however, we managed to get the trucks within a mile of the top by bucking drifts and dragging fallen trees off the road. Here we switched to five horses; three for the riders and two packhorses made up the train. The drifts were steep, but the horses dug in. They were used to this and knew what they were doing.

Teddy is an old hand; he is good with horses and knows how to pack. I carried my camera equipment and roll of maps; Roger had a portable radio. He was in a merry mood this cool, sunny morning and called over the radio as he dismounted and went off the air, "Horse 37, 10-7" (37 is the number of his car.) It was windy, so windy that the men had trouble in getting the shutters up, and their cowboy hats were endangered. It would be a short ride on the wind to that wide-open cirque and on

* The old wood range was replaced by an oil heater later in this season.

down to the world below, so they filled their sombreros with lava rocks and placed them behind a sheltering cliff. It was no loss of stature for Roger as it would take more than a sombrero to make a real cowboy of him; he is too neat and sharp in appearance. Teddy, the typical packer, would not be Teddy without that hat; it is part of his personality.

Because the season was cooler this year, there are not so many flies, but, remembering last year and the millions of flies that had met their death on the windows around the bed, I had an idea what might lie inside that box-like structure that holds the spring and mattress. I asked the men if they would tip it up and dump out the accumulation since it was too heavy for me to handle alone. Roger seemed amazed at what they found and said, "I don't blame you for wanting to have it cleaned out." This year I brought with me a large plastic sheet, large enough to cover the bed and hang down over the sides. Not only will it keep out the flies, it will also serve to keep the bedding dry in the driving rains that blow in around the windows.

While cleaning the bed, the men stood the springs and mattress on edge in front of the radio table, which was the only space available. Now who should call but Chiloquin air patrol with a message for relay to the Chiloquin district. I jumped up, draped myself over the mattress and springs; then, with my head hanging down and my feet in the air, I answered the call. I reached for the other mike and, in the excitement, mis-translated the message. Of course, anyone within sound of my voice would picture me in a more business-like pose and criticize my mistake.

The channel one radio they sent up this year is a rattle-trap.* It is full of pine needles, dried grass and dirt that keeps falling out on the table. It must have been dragged behind the pumper on a fire run. The squelch

* Channel one was used in communication with the three Winema districts. It was also monitored by a few of the Rogue River stations where they might be of assistance to the adjoining Klamath District of the Winema Forest. Channel two was monitored only by the Klamath District.

(static) cannot be tuned out or turned down. I can transmit, but cannot read the incoming messages without asking for many repeats. It will be a long summer if I have to endure this all day and night. Well, for the time being I'll click the switch on the contraption and call it my evening time off. It's been a long day.

As nighttime comes and darkness closes in around me, I feel a wave of loneliness. The lights in the valley twinkle silently. They look so near, yet out of sound and inaccessible, seem distant as the stars in my mood tonight. Can it be the wind that brings this lonely feeling? It has a mournful sound, almost a sigh of anguish in the tower eaves.

June 18th . . . The season opened with a bang. A severe lightning storm hit the district today. I got the azimuth and approximate distance on nineteen strikes within a few miles of the station. There were many I did not record as rain was falling in torrents. As the storm passed over the butte, I did not see definite strikes, but the cloud around me lit up in a flash of orange-pink. The flashes were almost constant. During this time, my hair was charged with electricity. It's a strange feeling.

June 19th. . . A message from Roger said they were making an attempt with the bulldozer to break through the one big drift now left. I can hear them down there in the fog. I must get the pin-curlers out of my hair; they've been in for several days.

Evening. . . The drone of the dozer has stopped. I have heard no more from the men. At 7 p.m. I went down to see what progress has been made. I crossed the snowfield in an open area. There, steeped in the murky light of dense, dark fog, I experienced an eerie sensation: no sense of direction, no sound but the crunch of my footsteps on crusted snow. It was spooky.

Snow still lies deep on the road down where the men were working this afternoon. They'll never break through that drift. It looks like I've combed my hair a month too soon.

June 20th. . . Jon and his family were up today. They came as far as they could by truck, then hiked, carrying the two small children and my mail. As glad as I was to see them, I crabbed most of the time they were here, griping to Jon about the equipment. It seems all I have to work with is cast-off junk the ground crews will not use. This channel one radio is driving me mad.

June 21st. . . I was up at 4 a.m. to turn on the heater — no gas; both tanks were empty. It was 34 degrees, but a fierce wind made it seem much colder. I dressed, putting on plenty of warm clothing, then waited until I thought the forestry men would be awake before making a call, for at this hour I would have to use the channel one which would be heard all over the Winema Forest, as well as Rustler, Brown and Dutchman peaks in the Rogue River Forest. (The district radios would be turned off at night.) With the radio reception so bad, it took many repeats before I understood that I was to start walking; a truck would meet me below the drift. I bolted down, cutting switchbacks where I could, to meet Lynn at Cold Springs Junction, five miles by road. Arriving at the guard station, I peeled off a few layers of clothing and had breakfast at a little cafe on the lake. What a treat this was for me, although Lynn was treading air in impatience with the delay my side-trip had cost in getting back up the mountain. Jon went on with the flatbed loaded with tanks of gas and a snowmobile, while I rode up in the truck with Lynn, crabbing much of the time about that radio. If I have to work with sheepherder equipment, I wish they would get me some sheep to herd. Lynn insists it's the same radio I had last year; well, it's not! Somehow, getting mad at Lynn or Jon was like getting mad at your own kids; you still like them.

After Jon had delivered the gas, he came down to meet us. I hopped on the snowmobile behind him and off we went, bounding over the snow through a heavy bank of clouds, leaving Lynn to plod his way on foot. My job may be monotonous at times, but speeding over a snow-packed mountain, clinging to the coattails of a forestry official added variety to this day. Oh, to be young again!

June 22nd. . . It is pouring rain, the wind is blowing like mad. Water slams against the window glass in waves. I baked bread and kept the water mopped up as it ran across the floor.

June 23rd. . . It stormed hard all night, leaving an inch or two of ice on everything and eight inches of ice on a single bolt that protrudes through a shutter. A pan of water in the tower cabin was frozen nearly solid during the night. The sun appeared occasionally today, reflecting a glaring light on thin fog, very hard on the eyes. The air is now a blizzard of frost or crystals of a frozen cloud; whatever it is, it's a chilling sight as it whirls past the tower on its way to nowhere.

Out of the storm came a voice from earth: "Is it snowing up there?" I recognized the voice and knew this was not an official call. From my point of view, or lack of view, it seemed a strange question to ask. Without taking time to consider the earthling's point of view, my answer came out, "I really don't know." (It seems I had taken my early instructions on radio procedures too seriously — "Make your conversations brief and to the point" — because sometime later I was told the caller had said at the time, "She's not very alert; doesn't even know if it's snowing.")

June 24th. . . Ice continues to build up; the ice on that bolt now measures eleven inches. Two of the boys hiked up, bringing with them the radio I had last year. Lynn, after hearing for himself what I had to work with when he brought up the butane, wasted little time in getting a better one up here. Well, the wheels made one turn; I can always hope for more. How wonderful it is to have a radio I can read and to be rid of that everlasting squelch. The boys also brought up eleven canteens of fresh water, carrying it on their backs. I would never have asked for water now; with all this ice around me and a full tank of gas, there is no shortage of water here. But oh, how good that fresh water is! I made coffee and the boys ate their lunch on the tower.

I walked down to the snowfields after signing off this evening. The trees reaching up through last winter's snow were heavily laden with ice and, fading into the fog, had an unhappy look. I saw large patches of red snow and thought it was blood; some poor creature has met its fate, I thought. A closer look showed it to be the red algae that grow in the snow at higher altitudes. They show up when the snow softens in spring.

June 25th. . . Several inches of frost formed on the sheets of ice last night. Wind-woven ruffles of ice are now coated with white, frosty lace. Little could be seen this morning beyond the frost-covered railing, for the mountain was wrapped in clouds. Subtle hues of a double rainbow hung in the fog as rays from the rising sun burned through. The arc was so near it seemed, almost, to be a halo for the tower — an eerie spectacle in the gray dawn.

Now, later in the day, the fog has lifted, revealing the landscape a-glitter with frost beneath a bright blue sky. What wonders of light and beauty must go unseen by the eyes of man! The "Glory Ring" is a phenomenon I sometimes observe when clouds are lying below the peak. Rainbow colors encircle the tip of the tower shadow as it reaches across the clouds. This phenomenon is created when the rising sun strikes the tower. Droplets of mist, arising from the vaporous clouds, act as tiny prisms, breaking the white light into its colorful components, thus forming the circle of colors around the tip of the shadow. In graceful motion, the colors change form and flow with the motion of rising mist. This may last for several minutes, following the angle of the sun. It slowly dissipates as the sun ascends the sky.

June 26th. . . The sun came up in a clear sky. Ice on the eastern side of the tower has begun to thaw, breaking loose from the shutters. Big chunks come rattling down, sometimes hitting the windows with a shattering bang as they clunk down on the catwalk or join the tinkle and clatter of breaking ice from the lower part of the structure as it loses its hold and thumps to the ground. Ice on the south shutter is beginning to melt and lose its hold, but re-freezes before it has time to fall. Odd figures now cling at strange angles to the ice-coated shutter.

The boys were up again today, bringing on their backs a pack of groceries that I had somehow overlooked when the horses brought my supplies up. I was happy to see the boys, but was sorry they made the extra trip. These were supplies that I needed, but could have done without. I may crab about the equipment here, but can't complain about anything else. My supervisors and the boys they employ are thoughtful and considerate of me in every way.

June 27th. . . It snowed several inches last night, drifting to eight inches in depth. A northeast wind is blowing snow up the crater wall; flying high above the rim, it gives the impression of a snowstorm downside up. Ice and frost are four to eight inches thick on everything. The open trapdoor is locked in ice against the cabin wall. Empty water cans on the catwalk are frozen together in an armor of ice, with long spears of ice jutting grotesquely in every direction from the wind-tossed freezing fog as it whirled around the tower last night. Nineteen inches of ice have now collected on that bolt. I do not know why this one bolt collects so much more ice than others, or how so much weight can hold onto such a small surface.

I was taking a batch of bread from the oven when I received a call from Lynn: "I'm sending Roger up to bring you down for a couple of days." Below, the moisture fell in the form of rain and fire danger is low. I filled my backpack with camera equipment, a good lunch, and necessities for two days' leave. Then, while pouring salt water on each step ahead, I slithered down with my backpack dragging on ice that protruded from the netwire, partially closing off the stairway.

I struck out down the road. No cut-offs for me today, for it was snowing hard and visibility was zero. I could easily lose my sense of direction in the storm. Below timberline, tree branches were bending from the weight of ice and snow. Gray and shadowless in the dim light and falling snow, they flanked the road as I strode over drifts to meet the truck.

"Merry Christmas." It was Roger's voice, although I could barely distinguish the end of his truck parked at a switchback below the drift. We bounded over rocks and mud down to the guard station, where I changed to summer clothing. The sun was shining brightly, a cool breeze fanned the warm air. It seemed incredible. I felt that I had fallen from a storm cloud onto the wrong planet.

July 3rd. . . Heavy smoke was moving in across the mountains from the north and west this morning. Drifting south, it darkened the sky and buried the earth. Visibility was near zero for much of the day. A special bulletin from the supervisory office reported the smoke to be from Alaskan fires that are burning out of control. It does not seem reasonable to me. Dutchman Peak called from near the California line to ask if I had information on its source.

July 4th. . . A temperature inversion this morning is holding the dense smoke below 7,500 feet. Unlike the clouds I see so often lying below me, this brown blanket is flat and motionless with only the sharp peaks of Mt. McLoughlin, Mt. Scott, and Pelican breaking through to the sunshine above where the air is bright and clear. It has been determined that much of the smoke is coming from field burns in the Willamette Valley. I share my private haven with the many nutcrackers that keep the air alive with their raucous calls.

July 5th. . . Gray clouds were streaming over the Cascade Crest when I signed off the air and went down for my evening walk. As I approached the tower on my return, I heard the call, "Pelican Butte, from Lake of the Woods."

"We have a four-year-old boy here on the campground. It seems he has been abandoned by some campers." The voice was from the fire guard on duty. I relayed the message to headquarters who in turn notified the sheriff. I was busy on relay between headquarters and Lake of the Woods Guard Station, during which time Lynn slipped in the words, "I wish you would get me a fire; I would know how to cope with that."

While we were trying to solve the little boy's problem, those massive clouds kept rolling in. Like billowing waves on a stormy sea, they covered the earth. The sun, now low in the sky, fired the stormy sea with crimson light. The lashing waves rose and fell across the wind-swept surface. Backlighted by the sun, the pink, translucent waves were edged in golden mist. A surging tide of luminous color swept over ridges and ravines from mountain top to the setting sun. Below the tower, an endless stream of gray fog was cascading over and down the crater wall as though it had a destination it must reach before the sun went down. As it dissipated before reaching the basin I could only wonder: What is all the rush about? Why this hurry to the end of time?

Cloud waves now washed over the peak, closing out the light for a time, then opened again as the sun, now fiery red and circled with rainbow colors, was sinking beyond the darkening clouds. This, I think, was the most glorious sunset I have ever experienced. Though still anchored to earth by the busy radio, I was above this radiant sea of clouds. It seemed, almost, that I was looking into the light of eternity.

The last I heard of the little boy, he was in the care of the sheriff. (Word reached me later that he had been with his parents and grandparents on a camping trip. When they broke camp, each thought he was in the car of the other couple.)

July 7th. . . I had company today, a nice surprise as the road is still drifted in. A couple who had been fishing a mountain lake came up to share their trout with me. "We have more in the truck," they said. "Since you like trout, we'll leave some buried in snow and mark the spot where you'll find them." After sign-off I walked down and, finding a branch sticking out of the snow, I reached down and pulled out a two-quart milk carton filled with fresh mountain trout. The little refrigerator is not working, but there is plenty of snow. I will feast for a day or two.

July 9th. . . Lightning storms have been keeping us on the alert for several days. They hit the Winema in late afternoon and continued well into the night. I could enjoy the ferocity of these storms were it not for fear of the dreadful results should a fire start and get out of control. I picked up one smoke on the side of Graylock, a little blue smoke on the near-side of a mountain — perfect, just like in the funny papers. I wish all fires were this easy to pin-point.

Lynn, Jon, and Roger made it around the big drift today after much work by men and truck. The road in this spot is still under the drift.

July 10th. . . Lynn and some of the crew were having trouble finding a small smoke that I had turned in. It was evidently a sleeper from a recent storm. It was barely visible when first spotted and went down before I could get the legal description. "I'll try to get you a bigger one next time, Lynn," I called.

"Not that; I'd rather have trouble finding a little one than try to get away from a big one."

July 13th. . . Roger came up last night. I returned with him to the guard station and brought my own car back. I do not feel so cut off from earth when I can look down and see my own car resting there. I spent the early morning hours, before signing on the air, bringing my supplies up and arranging space to stash them away or put them to use where needed.

July 14th. . . Rae came up with Roger today, bringing goodies from her store that she knew I would enjoy. Seldom do I have a woman to visit with, so this was a treat in more ways than one.

July 20th. . . This has been a quiet day. Lynn called late in the afternoon, "Turn on your transistor radio; the men are about to land on the moon." I listened while the Eagle landed. Then later in the evening, I went down to the point of rocks with my transistor radio and listened to the men as they stepped out onto the surface of the moon while, through the binoculars, I scanned that distant body through a clear, uncluttered sky. It looked so near; in fact, I felt as close to the moon as I did to the planet earth, for I could hear the voices of the men who walked the moon, but the voices of earth were silent to me, for all but the fire prevention crew had time off to celebrate and the forest radios were still. I was satisfied not to be one of the celebrants. How much more impressive it was, alone, on the point of rocks, away from all earthly noise.

As I watched the moon and listened to the voices, I tried to imagine what it must be like for these men of earth who were actually walking its surface. The feat seemed impossible, as impossible as — well, going to the moon, to quote a now-obsolete phrase.

July 21st. . . The flies are thick again; they splat against the window glass like big drops of rain. A few mosquitoes winged their way up from the lakes, but, on the brighter side, mountain bluebirds flash through the sunlight on radiant wings. The purple-red penstemons draped in and around the lava rocks are beautiful. They look too gentle and out of place on this rough peak. Last year there were only the noisy nutcrackers in the bird world, and the few blossoms that broke through the lava soil were scraggly and weather-beaten — sturdy bits of color on the rugged peak. Perhaps, because there was more snow last winter and now in blooming season there is little wind, the plants have a better chance to display their true identities. Though few in number, the penstemons, wild buckwheat and pussypaws are making the best of the season. There was no gentle beauty up here last year. There was only the beauty of wind-whipped clouds and the beauty in form and character of the timberline trees.

July 26th. . . Life on the mountain is rich in contrasts — you never know what the weather or airwaves will bring. It was near 7 p.m. when the call came from Cliff, fire guard at Lake of the Woods station: "A lady in her eighties has wandered into the guard station. She has been lost and wandering in the mountains since 11 o'clock this morning with nothing to eat. She became separated from her relatives on a camping trip and does not know where or who she is."

I relayed the message to the district office and they in turn called the sheriff. There had been no report of a lady missing. Well, what to do now? Poor Cliff; he did not know what he had on his hands. He was concerned for her welfare and, not knowing what else to do, he took her to the living quarters and gave her some coffee while waiting for further orders. One thing sure: she was in good hands with Cliff. He is a reliable and pleasant young man. However, this must have been a dilemma for him, for, like Lynn, he would know how to handle a forest fire, but the care and feeding of a little old lady would be somewhat out of his line.

It was not long before her relatives, who were camped nearby, came looking for her. She had been gone less than an hour; she was not lost, but had become confused and wandered away. No doubt it seemed to her that she had been out all day. I was happy to hear that she was reunited with her family.

This episode was barely over when I heard voices on the radio making threats. "OK, then, beat him up," was one transmission I clearly read. I knew by the voice that it was none of our boys, so I called the three guard stations, thinking some of them might be in trouble. No one had heard the call, so we decided it must be "skip." I had been bothered with this on Price's Peak, picking up conversations as far away as Louisiana. All this excitement cancelled my evening off the tower, but such is the life of the Pelican firewatch.

* * *

September 8th. . . Lightning risk has been high for some time, keeping the lookouts under constant tension. The prediction today was eighty percent chance, with all indications of a severe electrical storm. No storm developed, just clouds, but how welcome they are after two months of hot, dry weather and thick blue haze that hangs beneath the clouds during the threat of electrical storms. The blue haze creates false smoke and makes real smoke hard to spot.

I felt I had to get off this mountain for a little time of freedom. I asked for permission to leave for the evening to go down to the guard station and little store. Here I bought groceries that I do not need, just to visit with Rae for awhile.

It is always dark and usually windy when I return from these runs to the land of the free. With my arms full of bundles and a pack on my back, I stumble up the steps while the cold winds whirl and moan through the frame of the old tower. If I feel myself losing balance, I brush against the netwire and keep on going. Removing my pack, I raise the heavy trapdoor with my back, being careful not to slip or the door would slam down and smash my head between the top step and the door, as there is just a few inches between the two.

It was good to get back to my tower home. I re-read my letters and ate a quart of ice cream by the light of the "firefly," a gas light that gives off only a little more light, it seems, than the glow of a firefly. Since I get up at daylight and go to bed when it's dark, I need only ignite this whiff of gas on these occasions. I like to look out into the darkness, but I do not like the darkness looking in at me, as it seems to do when the flame is burning.

September 19th. . . I was climbing the steps in a rainstorm last night when some valued papers blew out from my pack and over the crater wall. There they were in the morning, forty feet below, glued to the wall by rain that had fallen in the night. I tied a rope to the leg of the tower for safety and let myself down to retrieve them.

October 1st. . . It is now October. There is a feeling of autumn in the air. Clouds are in the sky again and every dawn is a new experience in the play of light and color over earth and sky.

October 6th. . . This morning as the first light of dawn spread across the heavens it embraced a massive cloud that towered above the peak of Mt. McLaughlin. When touched by the light of dawn, the gray cloud tones turned to brilliant shades of red and gold, while the earth below lay deep in shadows for the sun had not yet risen. What a lovely birthday greeting! I'm sixty-six today.

* * *

It was mid-October when I received the call from Lynn: "From the weather report and the way it looks now, you had better come down today while you can. You could get snowed in up there and not be able to get your car down."

I was expecting the call, so I had everything packed and ready to go. I went "10-7" (off the air). Then, locking the trapdoor above my head, I jumped into the wagon and drove down ahead of the storm.

A day or two later the mountains were white with a foot of new snow. With two trucks, three men and myself, we set out to close the lookout and bring everything down for the winter. After reaching the point where the trucks could go no further, we all piled into the four-by-four. With all four wheels chained, we were able to drive close to the top, so we did not have far to carry supplies and equipment.

With mixed emotions, I watched as Jon loosened the wing-nuts and let the shutters swing down over the walls of glass. I was happy to be free of their oppressive confinement and the petty trials that are part of this work. Yet, through these walls I had witnessed rare beauty and phenomenal displays of light and atmosphere that I would see nowhere else.

I have felt the protection of these glass walls when ice and windstorms battered the peak and, on many a quiet, fog-bound night, I've lain, cozy and warm, wrapped in the dark of a cloud cocoon within these walls of glass on a mountaintop.

Chapter 5

Pelican Butte – 1970

It was Monday, June first when I received a phone call from Jon. "We would like you to be at the ranger station tomorrow, or Wednesday at the latest." This was two weeks earlier than previous years, but the weather had been hot and dry so I was prepared, except for last-minute packing.

I had a doctor's appointment for a checkup. When I returned later in the day to get a report on the tests, the nurse said, "The doctor wants to talk to you."

Well, what now? Is something seriously wrong? I waited in suspense for nearly an hour, rolling and unrolling the ragged pages of the waiting room magazines, wondering if that nurse would ever call my name. Then, "Alice Johnson?"

The nurse led me to the inner chambers and handed me one of those ghostly robes which I donned, then laid myself out on the papered slab.

Another thirty minutes dragged by. I kept thinking of that five-hour drive to Klamath Falls.

Finally my get-up-and-go got the best of me. I dressed and said to the nurse, "I am going to Klamath Falls; there is nothing the doctor can do for me now."

She seemed to feel it was important that the doctor see me, so again I undressed and mounted the slab.

Thirty-five miles' worth of time passed by; I was now counting time by the miles I had lost. I was making another attempt to escape when I thought I heard footsteps, so, back on the papered slab — just in time.

In came my doctor with that unhurried look that doctors always seem to present to their patients. He tapped my shoulder and said, "You should cut your weight down to 125 pounds." (I weighed 138.)

"I am large-boned; wouldn't that fact allow me to weigh more?" I pleaded.

"No, five-foot-five, 125."

So, was this all my frayed nerves had waited for?* I did not know then that I would lose more than half that excess weight by the time I reached the tower.

It was now 5:30 p.m. I headed straight from the doctor's office to Medford, then east to the Pelican Guard Station, while lightning split the night sky over northern California, and worked its way onto the Klamath District.

I expected to spend the night at the Pelican Guard. As a migrating firewatch, my sleeping bag and I would have rights to the lumpy hide-a-bed; however, there was no one at the station. Since I was not yet on duty, I did not have a key to open the door. The mosquitoes were too thick for me to sleep outside and I was too tired to drive on to Klamath Falls to rent a motel room. Besides, I did not want to leave the car with my summer's supply of groceries, clothing and equipment. It was now 10:30 p.m. and I had to get some rest. Since there was no room in the back of the station wagon, I wadded my well padded sleeping bag across the bucket seats in the front. Then, without undressing, I managed to wriggle myself down into it. I slept a few hours in spite of the fact that I was lying in two buckets with my body wrapped around a gear-shift lever in the cramped space of a compact car.

I unfolded my body at daybreak and drove into Klamath Falls, had breakfast, and was at the ranger station before it opened. There was the usual rush of people on opening day of the fire season, perhaps more excitement this morning because of the lightning storm. There had been many strikes on the district during the night. The fire control men had been out before dawn, checking the forest for smoke; there were bound to be a few fires.

* I later learned the reason for the doctor's concern. By seeing me, he could charge for the call. His name came up later as one of the doctors in the area who were taking improper advantage of Medicare.

Lynn said they were taking me up by snowmobile this year because the snow was soft and the drifts were too deep for the horses. This meant repacking, as I could take only the really necessary supplies.

Due to short notice of the change in plans, I worked hurriedly, going through the neatly packed boxes, picking out supplies while trying to decide what my needs would be. My only guidelines were Lynn's words "really necessary" and "as little as possible." There was no way of knowing the length of time for which I would need to prepare, nor did I know how much the snowmobiles could handle. After I had finished repacking, there were a few items that would not fit back into the boxes. These I dropped into my backpack to get them out of the way. I was to find out later that this was a lucky break.

The snowmobiles were towed on a flatbed behind one of the trucks as far as Cold Springs Junction. The five miles from there to the lookout is a low-gear grade under the best conditions. The road was now under heavy drifts. When the supplies were transferred to the smaller machines, I hopped on with one of the boys. Well, it looked like trouble was here and now. That mechanical beast bucked and "sunfished" over the lesser drifts and nose-dived into the bigger ones. I was not a good enough cowboy to ride it out, so I stepped off and watched the boys. They were working hard to gain headway. Many times they jumped off while the machines were going full speed, grabbing hold wherever they could, in an attempt to shove them over a drift. That gave those devilish things their chance: they swapped end for end and headed down drift with the boys in hot pursuit, catching up when the machines rammed into a drift or when the caterpillars clogged with soft snow. When this happened, the boys tipped them over to clean out the mush. The drifts were six, eight, or maybe ten feet high and accordion-pleated by the wind as it drifted snow through the trees. One thing was sure:

the little snow-bugs had no plan to cross those accordion drifts.

Being of no help to the boys, I suggested that I start out on foot, for I had to get to the lookout some way. The boys gave me a drink of their good cool water and I set out with the pack on my back.

I was now leaving the boys and their troubles and taking on a few of my own. With my boots, I kicked a toehold into the snow, clawed my way up a bank and slid down on the other side. It was one drift after another like this for a good part of the way. I plopped down a few times between drifts to rest and cool off, or leaned against a bank to let the snow melt on my back. In spite of the snow, the sun was hot and water was running from the drifts in streams. I finally dropped my pack, afraid I would never make it with that extra 32-pound weight.

At the first opportunity, I took a shortcut, following the edge of the crater wall, thinking it might have blown bare. Not so; a cornice curled over the rim. Had I not known the area well enough to stay away from the edge, the cornice might have caved off, dropping me over the crater wall. I knew how close I dared go by the trees I knew so well. My problem now was to avoid boulders and trees where the snow had melted underneath, leaving a treacherous shelf. With much effort I skirted them; I did not want this episode to end with the boys pulling me out of a tree-hole. As I walked near the lower branches, one leg would often break through the snow, leaving the other leg folded up somewhere around my ears. Branches sprang from their winter blanket, giving me a prickly swat and a spray of mushy snow. I was becoming sick from exhaustion. I knew I would have to make it to the top, but I did not know how just then. Taking the shortcut saved distance, but made the climb that much steeper.

My thoughts would go no further than the next step ahead, until I finally reached the tower. Here I rested on a stone step at the base of the tower and waited for the boys to arrive. They had abandoned the machines about a mile down and finished the trip on foot, carrying the most essential lookout equipment, my bedroll and the backpack I had dropped along the way.

While the boys were opening the lookout, I brought a ten-gallon water can down from the catwalk and filled it with snow, using the only thing available, a one-pound coffee can, for a scoop. The boys took it up to the cabin and set it on the heater to melt. Snow would be my source of water for the next five or six weeks.

Well, the initial problems were over; I now had time to think of the next one. My food supply was down in the drifts with the snowmobiles, somewhere along the trail. The only food on top was what I had dropped into the backpack when consolidating for the snowmobile trip. I searched the pack eagerly to see what I had: a can of baked beans, a package of graham crackers and a can of milk. At least I would have a fair diet as long as it would last.

After getting the radios set up, one of the boys called the district headquarters: "We've got to get supplies up here tomorrow; she only has food enough for tonight." The reply: "We'll send a plane out from Medford tomorrow and drop the supplies." The boys then went down. Once more I'd be alone until the sun took out the drifts.

The following day, as I was writing up the radio log, a plane circled the tower three times, making a drop with each circle. They were bringing my summer's supply of provisions that I had left in my car at the Pelican Station. One chute dropped onto a tree; one landed on two trees. Both canvas packs dropped to the ground. The third chute and canvas fell free. Several boxes of canned goods, home-canned in glass jars, landed safely without a jar broken. The other supplies, not as well packed, did not fare as well.

Groceries arrive on Pelican Butte by "special delivery." June 1970

One of the packs that had fallen from the tree fell in a leaning position against another tree, buckle-side down. It was too cumbersome and heavy to turn over, but with a lot of work I managed to open it from underneath. Boxes originally packed to go up by horse were now broken open, leaving everything in a jumble. Gravity took over as I loosened the strap and the whole glorious mess came tumbling out. Well marinated with vinegar, honey, cocoa and flour, it went rattling down the slope, picking up lava ash, leaf fragments, and pine needles en route.

Amid the wreckage of broken and unbroken supplies were my partial plates that I wear only on special occasions. The container had broken open, and there they lay, looking up at me from the mess in the big drop canvas as if to say, "Where on earth have you been all this time?"

A metal suitcase carrying a small tape recorder was demolished, a hole poked through one end by a branch from the tough old tree on which it fell. The recorder was undamaged. The only suitcase not damaged was the $11 case that had taken the twenty-foot free fall from the rope on Price's Peak. The one good, brand-name suitcase shattered like Humpty Dumpty.

I started big trouble when I opened the third canvas. It contained a broken bottle of liquid shampoo that drenched everything in the pack. I hauled snow up to the cabin in pans to let it melt on the heater, then tried washing shampoo off the supplies, but all I got for my work was bubbles. I hauled more snow, and more snow, but more water only meant bigger bubbles. The more I washed, the more the bubbles bubbled. It was a losing battle, like trying to put out a fire with gasoline. Everything was bubbling except me. I finally decided to use snow in the raw, but I don't think there was enough snow on that whole mountain to get rid of those everlasting bubbles. A few large baking potatoes were really in a lather. I wore out a lot of snow in getting them polished, but you never saw such clean potatoes in all your life.

I spent the rest of that day and the next morning from the first light of dawn until signing on the air at 9 a.m. in getting the supplies gathered up and cleaning them as well as I could with snow. I then hauled them up to the cabin for further sorting and cleaning, working on it in all my spare time. I dropped a few dispensable articles in the burn barrel, bubbles and all.

These were high-risk days for fires due to the electrical storm. Thunderheads were now building again, threatening another storm. I was alert at all times for fires while running up and down the steps carrying snow and supplies. At about 7 p.m., now totally exhausted, I did what no firewatch should ever do between the hours of dawn and dark: I lay down to rest. I started counting my blessings: how lucky that my clothes were in the $11 suitcase. Hand-tinted photos of my daughter and late husband were not damaged, the tape recorder was intact, and, oh yes, I still had my banquet teeth!

Well, I had not finished counting my blessings when a curse fell upon me that has brought many a firewatch down from a mountain forever. A sleeper from the last electrical storm started up on my district. It was spotted by a watchman more than twenty miles away. It was only eight miles from my station and building up fast. This is a total disgrace for a lookout; I was devastated. I am sure the other districts had plenty to say about that lookout on Pelican.*

Lynn evidently realized a little of what I had been through and could not have been more understanding. I was grateful for his consideration. It was one of my blessings that I had nice folks to work with. The boys who brought me up, or tried to, showed concern for me and were helpful in every way they could be under the circumstances.

After this exciting start of the season, life settled down to a more or less normal existence. I passed through the ice age, the age of flies and insects, and periods of high winds and lightning storms.

It was on the Fourth of July that some young friends, Jim, Lena, and their friend Kenny (not connected with the Forest Service), shovelled a cut through the big drift. The cut was higher and no wider than the jeep that broke through to be the first wheels on the peak this season. However, the jeep was not one you would see in a showroom window. It might better be described as a wonder of wheels and noise, propelled by Lena's faith in Jim's accomplishments and Jim's labor and ingenuity. Oh, I suppose it had an engine on it somewhere, but little else that you would notice.

As it was in other years when summer arrived and the roads became dry, the tourists began drifting up. I met many interesting people: writers, artists,

* Fortunately this fire was in an area of easy access. The crews were soon on it and had it out before it had time to spread.

photographers, hikers, mountain climbers and a deep-sea diver and undersea explorer, to name a few.

One clear, cool evening a photographer arrived. He asked permission to spend the night in his camper at the foot of the tower in order to be here for the sunrise. The following morning, as light was showing in the eastern sky, I looked out to see a robed statue poised on the edge of the crater wall, eyes fixed on the eastern horizon, a peaked nightcap listing slightly to the west in the cool morning breeze. The figure looked out of place in the line of ragged, storm-torn trees, natives of the mountain.

In due time the sun came up on its tour of duty with no display of grandeur. I called down to the gentleman, asking if he would like to come up and have some breakfast. He would, so up he came, robe, nightcap and all. He was draped in a fortune of camera equipment and shivering with the cold — colder, too, because he had not had heat in his camper last night. I admired his effort and was sorry he was unable to get any worthwhile pictures. One must live with the moods of this temperamental mountain to capture its moments of beauty, for beauty is borne on the winds and clouds. A scene, drab, without color under cloudless skies may vibrate with beauty in the wake of a storm when none but the lonely firewatch, by force of her work, is on hand to witness the rare and exciting display.

A couple from Medford was taking a birthday cake to friends who were camped on Four Mile Lake. Unable to find their friends' camp, they drove up to Pelican Butte and asked me to sit in for their friends and share the birthday cake. I put on the coffeepot, but before the "party" was over, a fire broke out on the Chiloquin district. There was confusion for a time as the Chiloquin lookout could not see the smoke I reported, and I could not see smoke on her legal description. Evidently there were two small fires. In the meantime, other tourists had arrived. It was hard to wade through them in that small space. However, I was glad they were there, for with their binoculars they could confirm what I was seeing or not seeing. Well, the story has a happy ending since the fires went nowhere and neither did that beautiful, delicious cake; this nice couple left it all for me.

Many of the tourists were from out of state, although most were from the immediate area of southern Oregon. I learned much about the country from old-timers who had rambled over those hills on foot or horseback or flown over by plane. They were familiar with territory that I could only picture in my mind as I related the topographic maps to the area as I saw it from this station.

On three separate occasions, when I had a day off, I went hiking with a resident of Klamath Falls. He was one of the oldtimers whom I had met on Pelican Butte. His historical and geographical knowledge of the area made the trips interesting. We climbed to the abandoned lookout on the summit of Mt. Scott, the highest point in Crater Lake National Park. We hiked to the lookout on Hogback Ridge and visited with the young attendant; and we explored the lava caves and other areas of natural and historical interest in northern California.

Tourists often brought up fresh fruits, vegetables, and sometimes sweets. How I appreciated this. It was good to have fresh produce to add to my diet of canned food. I appreciated the thoughtfulness of folks I had never met and am not likely to meet again.

In my six years as a lookout I never saw a fire of more than a few acres within sight of my station, other than the 640-acre grass fire that burned to the walks of the Oregon Technical Institute near Klamath Falls. One man was injured in this fire and one truck lost when it catapulted off Plum Hill. There were many small fires in my districts and in surrounding areas under other fire protection services that were also of concern to me.

A flat-car loaded with lumber caught fire and burned on the tracks, ignited by the hot box that starts so many fires along the railroad right-of-way.

A rash of fires was set by an arsonist. Usually striking at night, he set off a number of fires in close proximity. It is difficult, if not impossible, to give an exact legal description from the lookout at night, but night fires are not hard for the ground crews to spot.

I was often involved in relay work not associated with my work as a firewatcher. There were forest campground brawls to settle, a plane that crash-landed, car wrecks, car break-ins, and often lost hunters and hikers. I have received messages at all hours of the night on search or emergency, all requiring relay to the district office. It never failed to startle me when I was awakened by a voice in the night calling "Pelican Butte." I leaped to the radio, knowing with certainty it would be a call for someone in trouble. In a case of lost persons, I remained on the radio until the search was called off or the person found.

A young lady with an injured knee spent the night alone on Devil's Peak while her brother hiked to the headquarters of Crater Lake National Park for help. Although the forest was full of nice, young fellows, Teddy and his horses were dispatched to the rescue. The moral to this, boys: "Get a horse."

A lady hiker had taken one of the most difficult trails in the area, a rough trail of six miles with an elevation gain of nearly 4000 feet. When she did not return at a reasonable time, her relatives became frantic and called on the Forest Service for help. As is usual in such cases, it was too late in the day to send out searchers, but men were posted at trailheads throughout the night and relatives spelled each other. I remained on the radio until the wee hours.

We were very concerned for her safety since she was dressed only in shorts and T-shirt and the nights are cold in the mountains. We were all up early the next morning and men were on their way when she walked out on her own. There is nothing unusual about this — it happens often. The unusual thing, as I heard later, was that instead of being thankful for caring relatives and strangers who had concern and were glad to donate hours of the night for her safety, she was provoked that we had misjudged her capabilities and wasted our time. She told her would-be rescuers, "I can take care of myself." It seems she could, for she had spent that night among boulders that had retained some warmth from previous days of sunshine.

I heard little of the daily news unless it was of concern to me, as it was when an all-points bulletin came over the air warning everyone in the back country to be alert for a fugitive twin-engine plane, one of an international dope-smuggling ring. The plane was riddled with bullets and the pilot thought to be wounded and dangerous. En route from Baja, California, the plane had landed near midnight for refueling in Reno. The police had been alerted by the U. S. Customs officials. When the officers approached, the plane took off, without lights, in a hail of lead from the Reno officers' guns. The plane was tracked by radar and thought to be in our area, likely to attempt a landing in the back country. A firewatch on the Rogue River Forest reported hearing a plane flying low, without lights, over Crater Lake National Park. It was found the next day, abandoned on a landing strip near Beaver Marsh, about 50 miles northeast of the tower. It had nineteen bullet holes in the fuselage and wheels. I never heard what became of the pilot.

In late summer a spell of hot, dry weather arrived. The sky was bleached and the sun-scorched trees seemed lifeless. Each day added numbers to the burning index and the days seemed never-ending. I watched long hours through haze-filled atmosphere, often scanning the hills with heavy binoculars. My eyes were burning and my body was weary at the end of day, weary from monotony and the stress of being ever alert.

Evening after evening, when making late fire checks, I would watch the shadow of the mountain creep across the lowland, then merge with the twilight, the twilight into night, now a blessed period of rest and darkness until the sun again sent out its light across the cloudless sky.

During these drab days of this, my sixth season as a firewatch, I made a decision to break the ties that bound me to this work. When the rains came, bringing an end to the fire season, I would go down for the last time. Six years is not a long time in this work, but I longed to be free from my own sense of duty, for there were still many trails I wanted to follow, a few mountains I wanted to climb. I was now anxious to be on my way.

The fall rains came in a gentle mood, the patter of a few drops in the dust, followed by a steady downpour. Lynn called me to go down, saying he would want me back later if the weather cleared, to be on duty during the fall operation of burning slash. I laid off for a day with no sign of the weather clearing, then decided to go back on my own initiative to finish packing, for it was now late October and there was no telling what the weather might do. If the storm continued it would mean the end of the season.

The road was badly washed from the rain but, in keeping astride the ruts with wheels on rocks or sod, I arrived on top where it was now sifting snow on a light breeze. How cold and cheerless the cabin seemed; there was no warm feeling of "coming home." I was keenly aware of the storm's increasing intensity as I hurriedly wrapped the canned goods in paper and packed them in cardboard boxes ready to be taken down when the lookout was closed. I emptied the cans of water over the catwalk rail so they would not freeze and burst. The water splashed off in the wind and disappeared in the storm. I felt strangely alone in the hostile atmosphere. Snow now covered the ground and was drifting with lava sand — more of this and I knew I could have trouble in getting down.

I was on my own, for no one knew I was there. Intermingled with this apprehension was the anticipation of freedom and the sense of aloneness I always felt when going down for the season. Now was added a tinge of sadness in leaving forever this mountain where, in three summer seasons, I had experienced the equivalency of a lifetime with its storms and sunshine.

I was glad to be alone with my mixed emotions as the car crept slowly down the mountain road, breaking fresh tracks in the new-fallen snow.

A day or two later the sun appeared, revealing mountains white with snow. I returned to the Klamath Ranger Station to hear of Lynn's plans for bringing things down. Due to the storm, I knew he would be closing the station and I was glad I had everything packed, but I was dismayed when Lynn said, "We can't get your things down until spring; there's more than two feet of snow up there." I wasn't expecting this. "I must get my things down," I replied with fervor, knowing full well that next spring could only mean never, for the canned goods would freeze and burst the jars and the bedding would mold; besides, there were utensils I needed for everyday use. I pled my case in vain. Never have I seen Lynn more adamant. "There's no way anyone could get up there now," he reaffirmed. Two men had tried it the day before; it had taken them all day. After wading through snow to their waists, they returned, saying it would be absolutely impossible to bring anything down from there now. "You'll just have to wait 'til next spring." That was the final word.

I left the ranger station, stopping to bid good-bye to the friends who had opened the road last July. "Kenny and I can take you up — we'll get your things down," Jim said with full assurance, as though it were no great operation.

"You can't," I replied with Lynn's final words ringing in my ears.

Jim was backed by Lena, his ardent young wife. She assured me that Jim could do anything. Another argument lost — I felt suspended between rock walls, for I did not want to go against Lynn's will, yet I wanted desperately to rescue my things.

By some kind of luck, bad or good, I'd forgotten to turn in the forestry key. It took one day to get things ready for the trip and until noon the next day to complete preparations. The troupe consisted of Kenny, Jim, Lena, and Bitsy, their little black dog, and me. The equipment was Kenny's big truck, two snowmobiles, a large, heavy canvas, and other essentials. The snow was soft, enabling us to get the truck within five or six miles of the tower after dragging snow or shovelling it ahead of the truck. Here we unloaded the snowmobiles which were small machines, ridden astride like motorcycles. Lena and Bitsy rode behind Kenny. I jumped on behind Jim and clung to his jacket as off we went, like fleas from a hot griddle, on to the battle of Pelican Butte. The new-fallen snow, although drifted in places, was soft, unlike the steep, hard-packed drifts of June.

Although both of the boys were mechanics, they were used to heavier equipment and had never handled snowmobiles. We had a few pile-ups in the soft snow until I learned how to lean around switchbacks and swing with the action. Even then the ride had its ups and downs, with me going up as the machine went down and vice versa. My untimely landings made the steering much harder for Jim. The going was rough for the little machines seemed inclined to follow the line of least resistance, paying little attention to conventional steering. Jim used his legs to aid in braking and turning. It was a fight to the finish to keep the beasts on course. For sure, it was only the luck of the Irish that kept us from shooting over a bank or diving over the crater wall.

The machines came to dead end a half mile from our goal. Here Lena and Bitsy remained while the two men and I went on. It was now leg work and blind determination that enabled us to break through the two feet or more of soft snow before we could reach the lookout. There we used the canvas to wrap the heavy, boxed supplies, while I wrapped the bedroll and lighter things with the plastic bed cover. After the shutter was down and locked, I set out with my bundle sliding over the snow ahead of me.

My problem now was to keep the tow rope from tangling around trees or the bundle from slipping down over the wall. The boys, with their heavier load sinking into the snow, had all they could do to snake it down to the snowmobiles. Here we lashed a bundle behind each machine. With the bundles dragging behind on the snow, we set out on a wild ride back down to the truck.

This was Bitsy's first encounter with snow. Her body trembled with nervous excitement as she huddled in Lena's arms on the back of the snowmobile. All went well until that infernal machine hit a buried obstruction. The jolt was too much for Bitsy. She left Lena's arms in a flying leap, landing head first in a drift, leaving two little black hind legs kicking frantically above the pure white snow.

"Stop, stop!" Lena screamed, trying to make herself heard. The roar of the motor died down and Lena was off to the rescue. But Bitsy would have nothing to do with her now; she'd had enough of that outfit. Bitsy came bounding through snow to where Jim's machine had jumped trail, leaving us stuck in a drift. With pitiful eyes she pled with me to save her from those maniacs. I could offer no consolation: "Bitsy, don't look at me like that; I have all I can do to take care of myself." Left now with no other choice, Bitsy reconsidered her plight and returned to Lena's arms.

We made it down intact, just barely, for the canvas had worn through on the heavier load; only the boxes were holding its contents together. Had we any more distance to go, the supplies would have been scattered over the mountain again, as they were when dropped from the plane in the spring.

We were now racing the sun, not thinking of time by the hour until Jim looked down at his wrist — no watch — gone! Somewhere up those impossible miles an expensive watch was ticking off time in two feet of snow. Although I wondered how much of my summer's pay it would take to replace Jim's watch, my heart sank when Lena suggested they go back and find it, for I knew that would be impossible. They never could find it in all that snow, but I said nothing, for knowing Lena, she would try anything as long as Jim was behind her — and he always was. I was terribly concerned about getting down off that mountain; there was no time now to look for a "needle lost in a haystack" for it was already dusk in the lowland and night was closing in.

Kenny and I were left loading the bundles onto the truck while Jim and Lena back-tracked the trail. I was getting very worried and somewhat impatient when here they came, Jim and Lena prancing down the snowy trail, gleefully waving the watch. "We found it!" They had searched the areas where our tracks told we'd had trouble. Lena had spotted a tiny hole in the snow and, reaching down a hand's depth, had pulled out the watch.

It would surely seem now that the worst was over, but our troubles had just begun. The sun was down and the stars were out when the boys got the snowmobiles back on the truck. We were all very tired and ready to go, but the mountain was not ready to release us yet. The snow, so soft when we went up, was now frozen solid. Kenny had turned the truck while the snow was soft, good foresight on his part, for it left the truck headed downhill. He would not have been able to turn it now, as the wheels were down in icy ruts and the truck was riding a ridge of ice, endless ruts and an endless ridge, it seemed. Although the truck was chained and on a downhill grade, it wouldn't budge. Only by having the ice broken ahead of the truck could it be made to move at all. As Jim broke the ice, Kenny could make the truck lurch ahead a few feet; more ice broken, a few more jerks ahead and so on it went. I was unable to do a thing to help, but I felt very responsible. I felt guilty, too, for putting the boys through all this. Lena, Bitsy, and I shivered outside the truck until we saw it was making headway. We then piled in, remaining tense and silent while Kenny and the truck fought hard to get down that mountain road. Jim rode in the back with the snowmobiles when he was not out breaking ice. It was after 10 o'clock at night when we broke away from the mountain and the wheels of the truck rolled smoothly into Klamath Falls. After the rough rides of that day those last thirty miles were like sailing on the wings of an eagle.

Yes, Lynn was right; it was impossible. Lena was right: Jim, and here I must add Kenny, can do anything, including the impossible.

. . . "I would take nothing for the years I spent on the lookout, but am ready now to turn it over to someone younger. My best wishes to you and the Klamath crew."

With these words I closed a letter to Lynn telling him of our success, while omitting the frustrations of this, our final assault on Pelican Butte.

I enclosed the forestry key.

* * *